Twenty-first Century Drug Discovery

Edward D. Zanders

Twenty-first Century Drug Discovery

an Expanding Landscape

Springer

Edward D. Zanders
PharmaGuide Ltd.
Cambridge, UK

ISBN 978-3-031-93948-8 ISBN 978-3-031-93949-5 (eBook)
https://doi.org/10.1007/978-3-031-93949-5

This Springer imprint is published by the registered company Springer Nature Switzerland AG
The registered company address is: Gewerbestrasse 11, 6330 Cham, Switzerland

To Rosie

Preface

The twenty-first century has seen significant advancements in drug discovery, with technological innovations, new scientific insights, and evolving methodologies reshaping the landscape. Here are some key areas where drug discovery is expanding:

Personalized Treatments: With the advent of genomics and advances in molecular biology, researchers are now able to tailor drug treatments to individual genetic profiles, improving efficacy and minimizing adverse effects.

Targeted Therapies: Precision medicine involves identifying specific biomarkers that can predict how patients will respond to certain treatments. This has led to the development of therapies aimed at specific genetic mutations in diseases like cancer.

CRISPR and Gene Editing: CRISPR technology is revolutionizing the ability to edit genes, creating potential treatments for genetic disorders and other diseases at the molecular level.

Drug Discovery Automation: AI and machine learning are now being used to analyse vast datasets, predict how different compounds might interact with targets, and even design new molecules. This dramatically speeds up the initial phases of drug discovery, from target identification to hit discovery.

Predictive Modelling: AI algorithms can predict drug behaviour and toxicity, reducing the time and cost of development and increasing the chances of clinical success.

Drug Repurposing: AI has also been instrumental in identifying potential new uses for existing drugs, which can be faster and more cost-effective than developing new drugs from scratch.

Protein-based Therapies: The rise of biologics, such as monoclonal antibodies and gene therapies, has expanded the treatment options for conditions previously considered untreatable, including autoimmune diseases, cancers, and rare genetic disorders.

Cell and Gene Therapies: Innovations in cell and gene therapy have unlocked new possibilities, allowing researchers to target the root causes of diseases at the genetic and cellular level.

Immuno-Oncology: Immunotherapies, such as checkpoint inhibitors (e.g., pembrolizumab), have significantly altered the treatment landscape for cancer. By reprogramming the immune system to target cancer cells, these therapies offer a new paradigm in oncology.

CAR-T Cell Therapy: Chimeric Antigen Receptor T-cell (CAR-T) therapies are revolutionizing cancer treatment by using a patient's immune cells to target and kill cancer cells.

Stem Cells: The use of stem cells to regenerate damaged tissues and organs is a promising area of research that could lead to new treatments for degenerative diseases, spinal cord injuries, and heart conditions.

Nanotechnology: Nanomedicine and nanotechnology are making drug delivery more targeted and efficient, improving the bioavailability and reducing side effects. Nanoparticles can be engineered to deliver drugs directly to the site of disease, such as cancerous tumours.

Post-Marketing Surveillance: With the increasing use of digital health technologies, there is now a wealth of real-world data available to monitor the safety and efficacy of drugs post-market. This data helps refine treatments and uncover new therapeutic insights.

All of the above text was created by ChatGPT using the book title as input which vividly illustrates the power of large language models. From this point on however, the only words in the book come from the author's imperfect brain.[1]

The basic principles of drug discovery have remained more or less the same for many years, but the ways in which diseases are analysed and the types of interventions used to treat them have changed enormously, particularly over this first quarter of the new millennium. While clearly a good thing, it has nevertheless made my *How the Drug Discovery Industry Works* courses more difficult for me and my delegates. This is because I am restricted by the time available and therefore cannot do justice to the important new findings that are being reported on a regular basis. Rather than try to cram in too much material in a course, or to overextend the latter, I decided to cover these emerging themes in biopharmaceutical R&D in the form of a book, thereby allowing readers time to absorb the often complex and detailed material at their leisure.

This book is divided into two parts: the current state of drug discovery and development (as of 2025) and emerging opportunities, many of which were summarized by ChatGPT at the beginning. The earlier sections provide some background to the two main chapters that cover target discovery and interventions (i.e., medicines). Each chapter includes a list of references to help the reader discover more about the topic under discussion through reviews, primary publications, and websites.

[1] Assuming that that quoted publications were not produced using ChatGPT or equivalent as well.

Predicting the future of anything is fraught with hazards, but is perhaps more reliable with biopharmaceutical R&D. The multi-year timeframes for drug development mean that early discovery projects of the type described in the book may, after ten or so years, lead to successful medicines possibly with unique modes of action. Also, novel projects with real promise have for various reasons been abandoned only to be resurrected many years later with great success; antibody-drug conjugates and RNA-based drugs come to mind.

There are many aphorisms about predicting the future, but this advice by Peter Druker could be applied to all those connected with biopharmaceutical discovery:

The best way to predict the future is to create it.

Cambridge, UK Edward D. Zanders

Acknowledgements

This book grew out of my drug discovery training courses, and I am very grateful to the delegates who have been interested in the area and who have engaged in useful discussions. The courses would not be possible without the Oxford Biotechnology Network (OBN) and their helpful staff who have organized and hosted these over the years.

I thank Susanne Dathe from Springer Nature for agreeing to publish this book and Ramya Ravichandran who coordinated its production.

Thanks also to Patrick Byrne who led me in the direction of much-needed help with machine learning and deep neural networks.

I have tried to ensure that figures and related material provided by others have been properly acknowledged; I am grateful to Springer Nature, Elsevier, SAGE publishing, IQVIA as well as those with CC-BY-4.0 licenses for permission to use some of their material in figures and tables.

Also, I would like to acknowledge the *Cambridge Independent* newspaper in Cambridge England, whose uncompromising science coverage has alerted me to new information about companies and discoveries and the Cambridge Centre for Computing History for the timeline in Chap. 3.

Lastly and by no means least, I am grateful as always to my wife Rosie for her support and encouragement.

Competing Interests The author has no competing interests to declare that are relevant to the content of this manuscript.

Contents

Part I

Current State of Drug Discovery and Development

1 Introduction

Drug discovery and development has always been a complex business even when medicines were almost exclusively small organic molecules. At the time of writing, a quarter of a century has elapsed since the beginning of the new millennium. The once novel therapeutics, such as monoclonal antibodies, emerged in the last decade or so of the twentieth century and are now firmly established in the top sales ranks of branded prescription medicines worldwide. More recently, gene therapies and nucleic acid-based drugs have reached the clinic along with cell therapies that were once considered unthinkable as pharmaceutical products. There is great promise in these new approaches, but also great challenges to be overcome, particularly when trying to expand the use of these quite exotic technologies to treat the many millions of patients who suffer from common diseases. The following chapters review the current status of emerging approaches to drug discovery and development. Some of these will ultimately prove to be unviable, while hopefully many others will be successful and reduce the burden of disease and even be curative.

This chapter provides a brief overview of global disease trends, including mention of the COVID-19 pandemic. It will also summarise the changes in the nature of therapeutics that have emerged since the year 2000.

1.1 Disease Projections

The ultimate aim of biopharmaceutical R&D is to treat diseases, so any thoughts about what drug entities are going to be useful in the future must consider likely global disease trends over the following decades. Strong economic growth in countries with large populations, such as India, China and Indonesia, will inevitably lead to increased demands for treating the diseases currently associated with Western lifestyles. Also, global populations are ageing and require medicines to treat cancers, cardiovascular disease and chronic degenerative conditions. Lastly, rising temperatures will have an inevitable impact upon human health, particularly in

E. D. Zanders, *Twenty-first Century Drug Discovery*,
https://doi.org/10.1007/978-3-031-93949-5_1

poorer countries, with the possible emergence of new infectious agents with impacts similar to the COVID-19 pandemic.

1.1.1 Global Mortality Trends to 2060

Sleeman et al. (2019) have curated World Health Organization (WHO 2022) data to project disease burden up to the year 2060. Although their emphasis was on the future needs for palliative care in the very sick, these clearly relate to the most debilitating and life-threatening conditions that will require effective, globally available medicines. Their conclusions are sobering and briefly summarized as follows:

- 47% of all deaths globally (48 million people) will be related to serious health issues by 2060
- 83% of these deaths will occur in low-income and middle-income countries, which represents an 87% increase compared with 2016, due mainly to cancer deaths
- Suffering is projected to increase most rapidly among people with dementia.
- Serious health-related suffering is projected to increase most rapidly in people aged 70 years or older, with an increase among younger age groups, including children, as expected in low-income countries

The main disease types and morbidity/mortality estimates for 2016, 2030, 2045 and 2060 are illustrated in Fig. 1.1.
The data show a major increase in malignant neoplasms over the decades followed by significant increases in cardiovascular and lung diseases and dementia. Interestingly, the number of HIV-related deaths is projected to fall despite the challenge of producing an effective vaccine against this virus.

1.1.2 New Drivers of Excess Mortality

1.1.2.1 COVID-19

According to the WHO, the global excess mortality associated with COVID-19 was 14.91 million in the 24 months between 1 January 2020 and 31 December 2021, 9.49 million more than those directly attributable to COVID-19 (WHO COVID-19 2022). The pandemic has now run its course with the causative SARS-Cov-2 virus now becoming endemic like other well-established coronaviruses. The longer-term mortality figures are decreasing to reflect this, but there is still the lingering effect of *long COVID* which is poorly understood at the time of writing. Despite this, the impact of the pandemic on drug development that saw the rapid introduction of novel vaccines, funding initiatives, manufacturing scaleup and clinical trial design will help enormously when confronting other emergent viruses of concern.

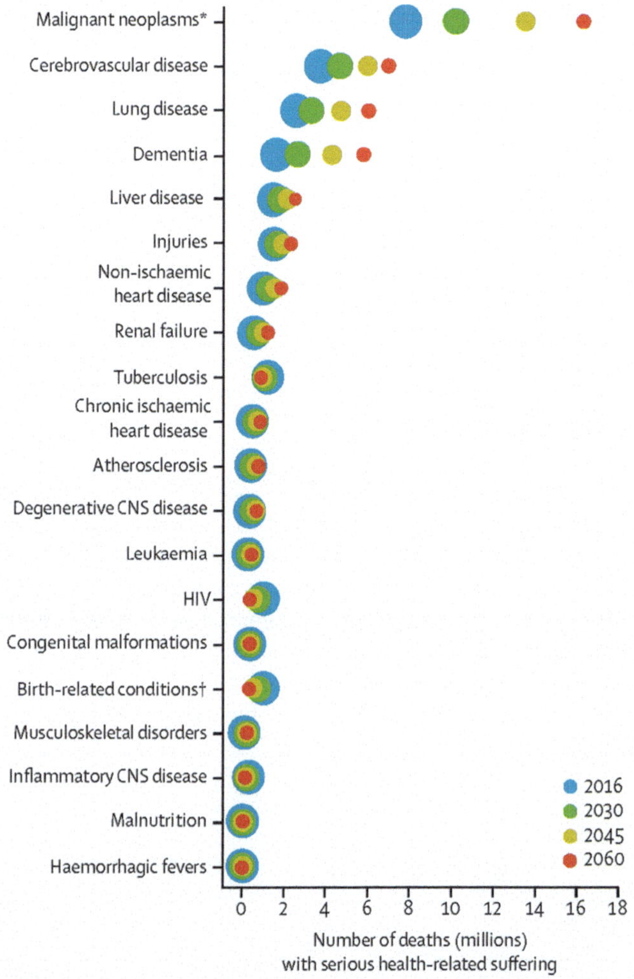

Fig. 1.1 Deaths in millions by disease area estimated for 2016, 2030, 2040 and 2060. (Figure courtesy of Sleeman et al. 2019 with permission)

1.1.2.2 Climate Change

To quote from the WHO's summary of climate change issues (WHO Climate Change 2023):

Research shows that 3.6 billion people already live in areas highly susceptible to climate change. Between 2030 and 2050, climate change is expected to cause approximately 250,000 additional deaths per year from undernutrition, malaria, diarrhoea and heat stress alone.

More detailed health outcomes may include:

Injury and mortality from extreme weather events
Heat-related illness
Respiratory illness
Water-borne diseases and related
Zoonoses
Vector-borne diseases
Malnutrition and food-borne diseases
Noncommunicable diseases
Effects on mental and psychosocial health

1.1.3 Ageing Populations

On 15 November 2022, the world's population reached 8 billion people and will take until approximately 2037 to reach 9 billion and 10.4 billion in 2100 (United Nations Global Issues 2025). While the rate of population growth is declining, the numbers speak for themselves in terms of future challenges for humanity. As always, effective medicines will be needed for a variety of ailments, notably the chronic diseases associated with ageing populations and infectious diseases developing in large, crowded populations. The issue of population ageing is illustrated in Fig. 1.2 using United Nations data for 2023 curated by the World Bank (World Bank Group Data 2025).

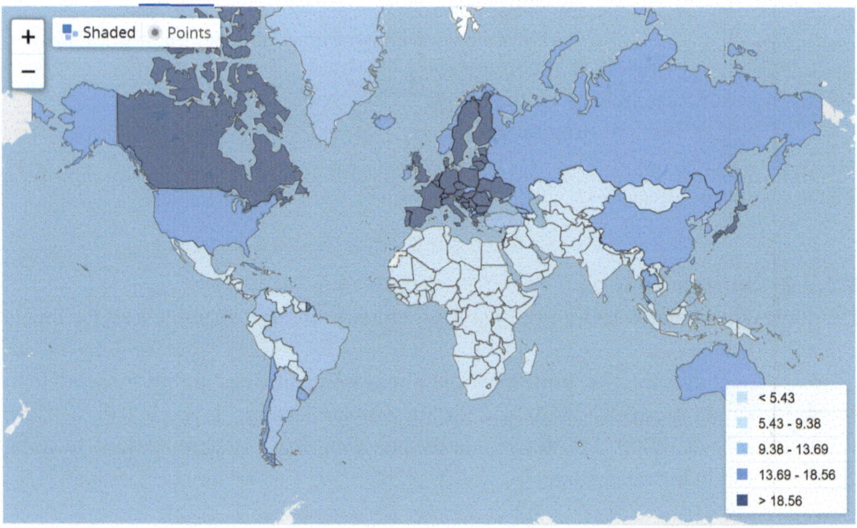

Fig. 1.2 Map of population density of over 65-year-olds (World Bank data under CC-BY-4.0)

The USA, although the largest market for pharmaceuticals, is only 36th in the world, having 16 percent of the total population being over 65, albeit with 52 million people in this group. China has 11.9 percent over 65 and India 6.1, but these still represent a market of 166.37 and 84.9 million people respectively.

The data presented above reinforce the idea that the market for treatments directed towards the chronic diseases of ageing will increase, with Japan being a dramatic example. The less developed world has, and will continue to be, a major healthcare challenge, however this is a snapshot in time; global economic developments later in the century may change the detail completely, for example by 2050, China and India could have surpassed the USA in gross domestic product. Whatever the changes in demographics, ageing populations will continue to require effective treatments for chronic diseases. Figures showing the greater proportion of medicine use by the elderly in the total population are available, but this fact was brought home to me during a visit to the British Museum in London. I came across a hall with very long tables filled with different medicines representing those that might have been taken by a man who lived to 76 and his wife, still alive at 78 (Fig. 1.3). The tables contained over 14,000 products, the estimated average for each person in the United Kingdom over their lifetime (which would have been over 40,000 if over-the-counter medicines were taken into consideration). This exhibition *Cradle to Grave*, was designed by Pharmacopoeia a group comprising a textile artist, video artist, and doctor (see Urban 75 2015). Both subjects received standard early life treatments such as vaccinations, used antibiotics, painkillers etc. and then specific medications for arthritis, asthma and breast cancer. To underscore the point of this section, the man died of a stroke having taken as many pills during the last 10 years of his life as in the first 66

Fig. 1.3 Part of *Cradle to Grave* art installation created by Susie Freeman, Liz Lee and David Critchley for The British Museum in London showing pills taken during the lifetime of a hypothetical UK-based man and woman

1.2 New Types of Medicine

In 2025, small organic molecules comprise the majority of prescription and over-the-counter medicines as has indeed been the case for many years. The reason is clear enough: they are generally easier and cheaper to manufacture at scale, often use an oral route of administration, thereby encouraging patient compliance, and can be replicated atom for atom as generic drugs. Of course it is not always possible to affect biological targets (mainly proteins) with small molecules, so larger entities like proteins have been used, insulin being a venerable example. The twentieth century saw the first use of cell transplantation (1957) and gene therapy (1990) both of which have developed significantly into viable therapeutic approaches for particular diseases. The result is a significant broadening of the pipelines of biopharmaceutical companies since the beginning of the millennium. This process should reasonably be expected to continue for years to come as the newer technologies become more established.

1.2.1 Changes from 2000 to the Present

Table 1.1 shows the top 10 products by global sales for the years 2003, 2013 and 2023 to highlight the changes in medicine type since the early 2000s. Small molecule drug brand names are in bold type, with the rest being protein therapeutics, mostly monoclonal antibodies (mAbs). Comirnaty is the vaccine against SARS-Cov-2 based on mRNA. There is a noteworthy change in the proportion of small molecules moving from 9 out of 10 in 2003, through to 5 out of 10 in 2013 and 2 out of 10 in 2023. The FDA date of approval is given for each entry as an illustration, recognising that in most cases there will be an equivalent for the EMA. Obviously, there is a significant lag time between initial development of a product, its approval and market position. This is illustrated by the top selling medicine in 2023 Keytruda (pembrolizumab) an antibody developed in 2003 by Organon and later passed on to Merck who filed an Investigational New Drug (IND) application in 2010. It received a Breakthrough Designation (BTD) from the FDA in 2013, initial approval in 2014 and approval for a wide range of cancers in 2017. Sales increased from 11.084 $Bn in 2019 to 25.0 $Bn in 2023. Since this book is concerned with how drug discovery and development might progress in the coming decades, it will be interesting to see if the top 10 sales of the future include not only small molecule and protein drugs, but also economically viable cell and gene therapies.

Table 1.1 Top 10 global medicine sales for years 2003, 2013 and 2023 (sources IMS Health (now IQVIA) and Verdin 2024 with permission). Bold text: small molecule drugs. FDA approval year taken from Drugs@FDA: FDA-Approved Drugs (2025)

Year	Brand name	Company	Indication	Sales $Bn	Yr of FDA approval
2003	**Lipitor**	Pfizer	Elevated cholesterol	10.3	1996
	Zocor	Merck	Elevated cholesterol	6.1	1991
	Zyprexa	Eli Lilly	Anti-psychotic	4.8	1996
	Norvasc	Pfizer	Hypertension	4.5	1992
	Epogen (Procrit)	Johnson & Johnson	Anaemia	4.0	1989
	Ogastro/ Prevacid	Takeda & Abbott Laboratories	Anti ulcerant	4.0	2003
	Nexium	AstraZeneca	Anti ulcerant	3.8	2001
	Plavix	Bristol-Myers Squibb and Sanofi-Synthelabo	Anticoagulant	3.7	1997
	Advair/ Seretide	GlaxoSmithKline	Respiratory	3.7	2000
	Zoloft	Pfizer	Anti-depressant	3.4	1991
2013	Humira	Abbvie	Arthritis	9.85	2002
	Advair/ Seretide	GSK	Respiratory	9.21	2000
	Crestor	Astra Zeneca	Elevated cholesterol	8.14	2003
	Lantus	Sanofi	Diabetes	7.94	2000
	Nexium	Astra Zeneca	Anti ulcerant	7.93	2001
	Abilify	BMS	Anti-psychotic	7.86	2002
	Remicade	J&J	Arthritis	7.83	1998
	Enbrel	Amgen	Arthritis	7.67	1998
	Cymbalta	Eli Lilly	Anxiolytic	6.46	2004
	MabThera	Genentech	Oncology	6.26	1997
2023	Keytruda	Merck & Co.	Oncology	25.0	2014
	Humira	Abbvie	Arthritis	14.4	2002
	Ozempic	Novo Nordisk	Metabolic	13.9	2017
	Eliquis	BMS/Pfizer	Anticoagulant	12.2	2012
	Biktarvy	Gilead Sciences	HIV	11.9	2018
	Dupixent	Sanofi/Regeneron	Allergy/ asthma	11.6	2017
	Comirnaty	Pfizer/BioNtec	COVID-19	11.2	2021
	Stelara	J&J	Psoriasis	10.9	2009
	Darzalex	J&J	Oncology	9.7	2015
	Opdivo	BMS/Ono	Oncology	9.0	2015

References

Drugs@FDA: FDA-Approved Drugs (2025). https://www.accessdata.fda.gov/scripts/cder/daf/index.cfm. Accessed 19 Mar 2025

Sleeman KE et al (2019) The escalating global burden of serious health-related suffering: projections to 2060 by world regions, age groups, and health conditions. Lancet Glob Health 7:e883–e892. https://doi.org/10.1016/S2214-109X(19)30172-X. Accessed 19 Mar 2025

UN Global Issues (2025). https://www.un.org/en/global-issues/population. Accessed 19 Mar 2025

Urban 75 (2015). https://www.urban75.org/blog/thousands-of-pills-cradle-to-grave-by-pharmacopoeia-at-the-british-museum/. Accessed 19 Mar 2025

Verdin P (2024) Top companies and drugs by sales in 2023. Nat Rev Drug Discov 23:240. https://doi.org/10.1038/d41573-024-00041-3. Accessed 19 Mar 2025

WHO (2022) WHO Global excess deaths associated with COVID-19, January 2020–December 2021. https://www.who.int/data/stories/global-excess-deaths-associated-with-covid-19-january-2020-december-2021. Accessed 19 Mar 2025

WHO Climate Change (2023). https://www.who.int/news-room/fact-sheets/detail/climate-change-and-health. Accessed 19 Mar 2025

World Bank Group Data (2025). https://data.worldbank.org/indicator/SP.POP.65UP.TO.ZS?end=2023&start=1960&type=shaded&view=map. Accessed 19 Mar 2025

Overview of Current Drug Discovery and Development

2

2.1 Introduction

This chapter provides the background for the subsequent more detailed coverage of science and technology being applied to drug target discovery and to the potential medicines arising from them. It summarises the basic processes currently used in drug discovery and development starting with the commercial and regulatory background followed by a description of different elements of the drug discovery pipeline. A more detailed coverage of these topics can be found elsewhere, and in the second edition of my book The Science and Business of Drug Discovery (Zanders 2020).

2.2 Commercial and Regulatory Aspects of the Biopharmaceutical Industry

2.2.1 Market Size

The following table (2.1) is based on projections of global pharmaceutical sales by region (IQVIA Quarterly Pharmaceutical Market Outlook 2023). The EU4 market consists of Germany, France, Italy and Spain. Pharmerging countries include Brazil, Russia, India, Turkey and Mexico.

The USA continues to be the largest market and is likely to stay that way for some years. However, the projections for economic growth by country shown in Chap. 1 imply that the USA's dominance in the pharmaceutical market may not last for ever, with China being an obvious successor in the future. Japan is a large market with an ageing population and has a very productive research culture, yet the domestic BioPharma scene is not as developed as in China, for example (see Kessel and Vickrey (2024). To place the \$2.0Tn value of biopharma within the context of other industries, in 2023, global music revenues were around \$28Bn, advertising spending around \$730Bn and global automobile revenues around \$2.5Tn.

E. D. Zanders, *Twenty-first Century Drug Discovery*,
https://doi.org/10.1007/978-3-031-93949-5_2

Table 2.1 Pharmaceutical sales projections

Country/region	2022–2027 share of growth ($Bn)	2027 market share %[a]
USA	297.96	46
EU + UK	74.49	14
China	34.38	9
Japan	0	4
Tier 2 + 3	91.68	12
Rest of world	80.22	15

Data courtesy of IQVIA (2024)
[a]Estimated total market in 2027 $2.0Tn

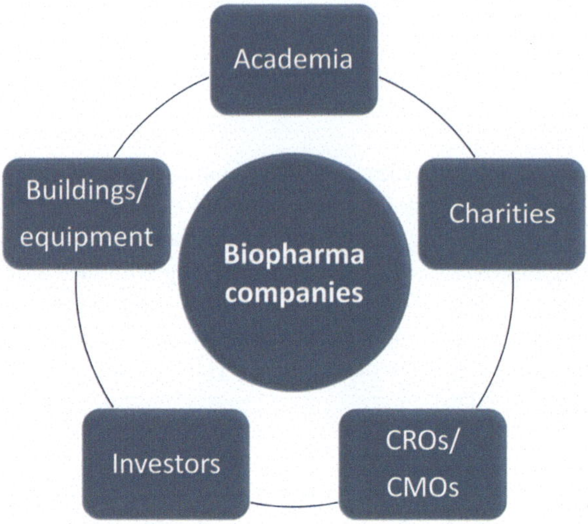

Fig. 2.1 Main components of the drug discovery/development network

2.2.2 Industry Organisation

The global biopharmaceutical industry consists of a large network of interacting organisations supporting the discovery and development of drug products by companies ranging from *big pharma* to small biotechs. A top-level view of this network is illustrated in Fig. 2.1.[1]

Each of the above elements is described in the following sections:

2.2.2.1 Biopharmaceutical Companies

According to data from Statista (2024) there were 6124 biopharmaceutical companies with active R&D pipelines in 2024. Table 2.2 shows a list of the top 10 biopharmaceutical companies by 2024 sales.

[1]Underpinning all these activities are the regulatory agencies that will be covered later

Table 2.2 Top 10 major biopharmaceutical companies by sales

Company	Sales $Bn
Johnson & Johnson	53.5
AbbVie	52.7
Novartis	52.5
Merck & Co	50.8
Roche	49.1
Pfizer	48.2
BristolMyersSquibb	44.4
AstraZeneca	43.8
Sanofi	40.8
GlaxoSmithKline	36.8

Figures from Verdin (2024) with permission

While each of the above companies has the resources necessary to discover and develop medicines for multiple indications, there are many more smaller organisations ranging from medium-sized companies to small startups spun out of universities. According to a report by IQVIA (2022), in 2021, big pharma was responsible for 19% of drugs under development but represent less than 1% of all companies involved in R&D. Despite this, they generate by far the largest sales revenues as summarised in the following market breakdown:

Large pharma: >$10Bn
Mid-sized pharma $5Bn-$10Bn
Small pharma $500 M-$5Bn
Emerging biopharma <$200 M

The Emerging Biopharma (EBP) segment is of particular interest compared with the other sectors as it is responsible for 65% of the total drug development pipeline, with an additional 7% being developed in partnership with larger companies. Big Pharma will boost their pipelines by acquiring assets from EBPs or indeed buy the whole company. The global contribution of EBPs to drug pipelines is broken down into approximate values of 50% for the USA, 20% for Europe, 17% for China, 6% for South Korea and 2% for Japan (source IQVIA 2022).

The collaboration between the German immunotherapy company BioNTec and Pfizer Inc. is a notable example of EBP-Big Pharma collaboration. The messenger RNA technology developed in the smaller company could be used to create an effective vaccine against SARS-Cov-2 and developed on a global scale using the resources available to Pfizer. As well as saving many lives during the COVID-19 pandemic and stimulating intense interest in mRNA technology, the arrangement proved very lucrative for Pfizer when demand for the Comirnaty vaccine was at its greatest.

2.2.2.2 Contract Research and Manufacturing Organisations

Contract Research Organisations (CROs) have been in existence since the 1930s and support biopharmaceutical R&D on a pay for service arrangement providing an economical alternative to in-house provision of expensive development activities.

According to a report by Global Market Insights (2023), the 2022 market size was $56.5Bn with a projected increase to $108Bn in 2032. The drivers of this growth include the growing outsourcing of R&D, increased R&D spending, more clinical trials in Pharmerging countries and technological advances. Clinical research services represent the majority of outsourcing activity, with projected revenues of $63.55Bn by 2032. Other activities include early phase development, laboratory services and regulatory consulting. Some of the major CROs with global presence are listed here:

ICON
IQVIA
Parexel
Syneos Health
Fortrea
PPD
Medpace
Charles River Laboratories
CTI

Contract Manufacturing Organisations (CMOs) are particularly important for providing medicines, medical devices and diagnostics at scale. The components of the manufacturing process that can be addressed by CMOs include production of small-scale batches for method development and stability analysis, large scale production, and adherence to the stringent regulatory requirements needed for products destined for use in humans. The CMO market size was $171.18Bn in 2024 with a projected rise to $230.38Bn in 2029 (Mordor Intelligence 2024, 2025). Contract Development and Manufacturing Organisations (CDMOs) as the name implies, provide integrated development and manufacturing services to biopharmaceutical companies and can be of particular use to companies without the resources and expertise to conduct a full development programme.

Fill Finish

One of the most impressive aspects of drug manufacture in recent years has been the scaleup of vaccine production in response to the COVID-19 pandemic. Both viral and mRNA vaccines are biological products generated in bioreactors and subjected to a series of polishing steps to remove impurities. The sterile products are then subjected to aseptic fill-finish where they are transferred by needle to a sterile vial or prefilled syringe and then closed. All of this requires sophisticated engineering and considerable resources of raw materials. The German company SCHOTT for example, needed thousands of tonnes of glass to manufacture more than 50 billion vials, an estimated 2 billion of which were assigned to the COVID-19 vaccine campaign in

2021 (Ramakanth et al. 2021). The market share of biotherapeutics is likely to increase over the years thus increasing demand for advanced manufacturing technologies and the associated raw materials (BR Business Research Insights 2024).

2.2.2.3 Academia

In July 1945, Vannevar Bush the Director of the Office of Scientific Research and Development in the USA wrote *basic research is the pacemaker of technological progress* in his influential report Science-the Endless Frontier.[2] To quote France A. Córdova in the 75th Anniversary edition:

> *He advocated for government funding of basic research in universities, colleges, and research institutes because that's where the talent was. He admired the freedom of inquiry he experienced as an administrator at MIT. Bush lauded curiosity-driven research with people working on "subjects of their own choice".*

This report for President Harry S. Truman eventually led to establishment of the National Science Foundation (NSF). In terms of biomedical research, the NSF has focused on basic mechanisms and technologies, while the more venerable[3] National Institutes of Health (NIH 2025) are closely aligned to the development of new medicines.

The foundations of drug discovery and development lie in basic research mostly undertaken in academic laboratories. All of the component disciplines such as medicine, biology, chemistry, engineering and computer science provide the knowledge for making discoveries and the tools for turning these into products. A significant number of drugs and technologies have originated in academic research laboratories; anticancer compounds based on cis-platin and CRISPR-Cas9 gene editing being just two examples. However, this is not the basic remit of an academic institution, one that is dedicated to fundamental research that can be freely disseminated in the public domain. This situation creates the tension arising between those conducting *blue sky research* in universities and those in the biopharmaceutical industry who apply science in a highly controlled environment. While this tension will never disappear, it at least has been ameliorated since Bush's time in a number of ways. These include commercialisation of publicly funded research in universities and the creation of links between academia and industry dedicated to using the best of both worlds. The article *Drug Discovery in Academia- the third way?* (Frearson and Wyatt 2010) provides a useful overview of academic-industry collaborations and suggestions for how these may be enhanced to make drug discovery more efficient and productive. Table 2.3 lists a few of the global academic centres specifically established for drug discovery/development collaborations.

[2]Covered in Science the Endless Frontier 75th Anniversary Edition (2020)

[3]See NIH History (2025)

Table 2.3 Examples of dedicated drug discovery centres

	Organisation	Location
North America	Broad institute	USA
	Small Molecule discovery center (SMDC)	USA
	University of Pittsburgh Centre drug discovery institute	USA
	University of British Columbia—Centre for Drug Research and Development (UBC-CDRD)	Canada
Europe	LifeArc	UK
	University of Dundee—Drug discovery unit (DDU)	UK
	University of Leuven—Centre for Drug Design and Discovery (CD3)	Belgium
	The Lead discovery center GmbH (LDC)	Germany
	Drug discovery and development institute IMS	France
Asia/ Pacific	Institut Pasteur Korea	South Korea
	Eskitis institute for cell and molecular therapies	Australia
	Peking university Institute for Drug Discovery and Development (IDDD)	China
	Centre for drug discovery and development	India

2.2.2.4 Philanthropic Organisations

The above title represents a diverse range of organisations dedicated to treating diseases which are summarised in the following sections:

Charities

To quote from a report by the Fraser of Allander Institute of Strathclyde University in Scotland (Fraser of Allander Institute 2021) *In the UK, there are an estimated 166,000 charities operating for a wide variety of causes across society, generating turnover of around £48Bn annually.* Medical research spending by UK charities in 2019 was £1.8Bn in 2019 which represents about 14% of all UK health-related research funding. Without this funding the UK Government and public bodies would need to increase their direct funding of health-related research by around 85% to cover the shortfall. Prominent UK-based charities include the Wellcome Trust, Cancer Research UK, The British Heart Foundation, the Alzheimer's Society and the Great Ormond Street Hospital Charity.

Non-profit Organisations

Charities make a significant contribution to drug discovery and development in different ways; for example they support university research into disease mechanisms, or the clinical evaluation of drug candidates that may not fit in with the remit of the biopharmaceutical industry. Sitting loosely within the category of charities, dedicated non-profit organisations have been established to find treatments for rare diseases, for example the Medicines for Malaria Venture (MMV).

Precompetitive collaborations between for-profit and non-profit organisations include the *Open Targets Collaboration* involving EMBL-EBI, Genentech, MSD,

GSK, Pfizer, Sanofi, and the Wellcome Sanger Institute among others with the objective of identifying novel targets for drug development.

2.2.2.5 Investment Community

Drug discovery and development is an expensive and high-risk enterprise. There is no certainty of success, and the time from early funding to a return on investment can be many years. Despite this, the potential rewards for success are great, both financially and in terms of human benefit.

Charities have already been mentioned as important supporters of R&D into new medicines. Their resources come from donations and investments in the financial markets with significant funding coming from investor groups: these include individual angel investors, venture capitalists (VCs) and venture capital trusts (VCTs) and biopharmaceutical companies can themselves invest through dedicated venture arms. Venture capital started in its current form in 1946 when Georges Doriot from Harvard Business School raised a $3.58 million fund to invest in companies that commercialized technologies developed during WWII. His initial $200,000 investment turned into $1.8 million in 1955 when the company went public (Investopedia 2024). Since then, VC investments helped to drive the IT industry and the early biotechnology companies in the 1970s. Genentech, now part of the Roche Group, has a bronze statue at its headquarters depicting two people sitting around a table at a San Francisco bar. They are Robert A. Swanson a young VC and Dr. Herbert W. Boyer a university scientist who helped to develop recombinant DNA technology. Boyer initially wanted to give Swanson 10 minutes of his time, but the meeting lasted 3 hours and lead to the foundation of Genentech. Thereafter the attractiveness of biotech to funders grew to the level it is today.

A VC investment can be made at different stages of drug development, the earlier being riskier due to possible attrition later on. The VC (or VCs) will expect to make a profit through an exit; this could be after the investee company is merged with another, acquired, often by a larger biopharma, or floated on the Stock Market in an Initial Public Offering (IPO). In the worst-case scenario, the company is sold in a *fire sale* with a loss on the investment. Investments can be made at different stages as follows:

Pre-Seed: Moving from basic idea to business plan
Seed Funding: VC or other funds to support product development
Early-Stage Funding: Capital for increasing product development and sales
Further funding rounds: denoted as Series A, Series B, Series C and occasionally Series D. Each round involves an increase of capital investment (in the millions) and provision of stocks/shares to investors as well as board involvement

From the perspective of a VC, biotech investment is resource intensive and fraught with risk. It is therefore advisable for VC organisations to have a portfolio of investments in the hope that a small number become successful to compensate for the inevitable failures. This is illustrated by comments made by Marc Andreessen, head of large VC firm Andreessen Horowitz (described in an article for Corporate

Finance Institute (CFI) 2025). His company review about 3000 startup proposals per year, i.e. approximately 12 opportunities per day (50 weeks per year, 5 days per week). They look at 200 very seriously and invest in about 20 annually, of which 15 will generate nearly all of the economic return. If the odds of being funded are 0.7%, and the odds of a funded company succeeding are 8%, the total odds of success are 0.05% (1 in 2000).

Venture Philanthropy
Venture philanthropy is VC funding directed at philanthropic organisations. As well as providing capital, such VCs will offer hands-on expertise to help the organisation develop. For example, healthcare-related investors can use biopharma executives to direct programmes. An early exponent of this practice was John D. Rockefeller III who coined the term *venture philosophy* in 1969 and his Rockefeller Foundation remains a leading investor in philanthropic enterprises. Subsequent dotcom billionaires have continued the practice, establishing for example, The Bill and Melinda Gates Foundation and The Chan Zuckerberg Initiative (CZI). A breakdown of the CZI's so called *biohubs* gives an idea of topics considered to be important in progressing healthcare. These include advanced biological imaging, bioengineering circulating immune cells for early disease detection and wearable microfluidic-based monitors enabling real-time analysis of inflammatory signalling.

2.2.2.6 Legal Support

The need for legal input into the activities of the biopharmaceutical industry will come as no surprise given the central importance of intellectual property law for its commercial activities.[4] In addition, legal expertise is required for managing the contracts and agreements undertaken for company licensing deals, collaborations and other activities within the industry. Human subjects are exposed to potentially harmful experimental medicines which can have legal consequences. For example, class action lawsuits have been made against GSK, the originator of Zantac which was contaminated with an impurity known to cause cancer. This has obvious implications for the consumers affected and the company itself and highlights the importance of legal expertise in supporting commercial activity.

2.2.2.7 Equipment and Consumables

All aspects of the drug discovery and development process require specialised equipment and consumables. A thriving industry has been built up to supply instruments, laboratory supplies and reagents, including radioactive molecules. Consumables include pipettes, sample tubes, tissue culture flasks, disposable gloves and test kits. The use of disposables increased dramatically during the COVID-19 pandemic which raises the question of how to deal with the large amount of plastic waste. Although the industry is making efforts to overcome the problem there is no easy solution given the nature of scientific R&D with its potentially hazardous

[4] Pharmaceutical patents are covered later

materials and the need for sterile materials. The market size for equipment and consumables is significant given that an estimated total R&D spend by the biopharmaceutical industry in 2024 will be over $213Bn. Key players include: GE Healthcare, Bio-Rad, Agilent technologies, PerkinElmer and Thermo Fisher Scientific. Some of these companies have formed from the mergers and acquisitions of smaller organisations. GE Healthcare, for example acquired Pharmacia AB, a Swedish company that used gels of cross-linked dextran to separate protein molecules on the basis of molecular size. These Sephadex® products were developed in 1959 originating through early research on protein purification at the University of Uppsala by Nobel laureate Arne Tiselius, among others. Sephadex® products and their variants are still the mainstays of the protein and nucleic acid purification techniques used for researching and manufacturing biotech medicines.

2.2.2.8 Buildings and Infrastructure

The major biopharma AstraZeneca (AZ) spent $1.3bn on its R&D facility in the Cambridge Biomedical Campus in the United Kingdom, giving some idea of the costs involved in providing infrastructure for the industry. The large figure quoted was due to the complexity of the design, building cost increases, environmental considerations and the cost of new technologies. In this AZ example, Cambridge was chosen because of its strong university, biotech and hospital links.

2.2.3 Regulatory Affairs

The primary products of biopharmaceutical R&D are medicines, medical devices and diagnostics that impact upon human health and therefore must be strictly regulated. The agencies that issue regulations for the major markets, notably the USA, Europe, Japan, and China, exert the most influence on global drug development. The introduction of drug (or related) products lies with *competent authorities* i.e. national agencies with relevant legal powers. The major regulators and some competent authorities are listed in Table 2.4.

2.2.3.1 Selected Regulatory Agencies

The Food and Drug Administration (FDA)[5]

The FDA was created in the USA as part of the Food Drug and Cosmetic Act signed into law by President Roosevelt in 1938. Amendments introduced later include: The Kefauver-Harris Amendments of 1962 which strengthened the requirements to prove the safety and effectiveness of drugs and The Medical Device Amendments Act of 1976, for medical devices. The reauthorisation and modernisation of The Prescription Drug User Fee Act (PDUFA) and the Medical Device User Fee and Modernization Act (MDUFMA) was undertaken in 2007 plus the Best Pharmaceuticals for Children Act (BPCA) and the Pediatric Research Equity Act

[5] See Food and Drug Administration (FDA) for full coverage of the agency's activities

Table 2.4 The main regulatory agencies and some national competent authorities

Agency	Location
Food and Drug Administration (FDA 2025)	USA
European Medicines Agency (EMA 2025)	Netherlands
Pharmaceuticals and medical devices agency (PDMA)	Japan
National medical products administration (NMPA)	China
Therapeutic goods administration (TGA)	Australia
Agência Nacional de Vigilância Sanitária (ANVISA)	Brazil
Health Canada	Canada
Central drugs standards control organization (CDSCO)	India
Ministry of food and drug safety (MFDS)	South Korea
Medicines and healthcare products regulatory agency (MHRA)	United Kingdom
L'Agence nationale de sécurité du médicament et des produits de santé (ANSM)	France
Bundesinstitut für Arzneimittel und Medizinprodukte (BfArM) Paul-Ehrlich-Institut (PEI)	Germany

(PREA). Lastly, the twenty-first Century Cures Act was introduced in 2016 to accelerate the introduction of new medical products by improving clinical trials, evidence gathering and use of novel therapeutics.[6] For example, a therapy may have a RMAT (regenerative medicine advanced therapy) designation; many of these will be gene or cell-based and therefore covered in this book at some length. A summary of the FDA's areas of interest in drug regulation can be found in: FDA Focus Areas of Regulatory Science Report (2022).

The parts of the FDA that are most relevant for biopharmaceutical development are the Centers for Drug Evaluation and Research (CDER) and Biologics Evaluation and Research (CBER). They cover the key regulatory applications required for the marketing approval of NMEs (new molecular entities) and biologics (proteins, nucleic acids etc). The Investigational New Drug (IND) application is for administering the drug candidate to humans for the first time and the New Drug Application (NDA) or Biological License Application (BLA) for marketing approval.

The European Medicines Agency (EMA)[7]

The European Medicines Agency, based in Amsterdam, was established in 1995 to coordinate a network of national agencies from all European Union (EU) countries and their political bodies. The EMA is organised into seven scientific committees of which the Committee for Medicinal Products for Human Use (CHMP) and The Committee for Advanced Therapies (CAT) are most relevant to this book. The latter committee is involved with the regulation of advanced therapy medicinal products

[6] Part of the FDA's Critical Path Initiative

[7] See European Medicines Agency

(ATMPs) such as cell and gene therapies. As with other agencies, the EMA is concerned with improving the efficiency of the drug development process, so it has published the EMA Regulatory Science Strategy to 2025 (2020) that mirrors the FDA's critical path initiative.

The EMA equivalents of the FDA's IND and NDA applications are the Clinical Trial Authorisation (CTA) and Marketing Authorisation Application (MAA) respectively.

Japan's Pharmaceuticals and Medical Devices Agency (PDMA)

The Ministry of Health, Labour and Welfare (MHLW) is run by a minister in the Japanese government and consists of different bureaux, including the PMDA (Pharmaceuticals and Medical Devices Agency) established in 2004 for drug development authorization.[8] As well as regulating conventional drug substances, the PDMA also covers so-called Regenerative Medical Products, i.e. cell and gene therapies.

China's National Medical Products Administration (NMPA)

The NMPA (NMPA 2020) emerged in 2017 from a reorganization of Chinese regulatory agencies. Given the significant increase in drug development by Chinese companies its penetration of global markets, the NMPA is an important national regulatory agency. The organisational divisions for drug regulation are similar to those of the FDA and EMA, (Division of Chemical Drugs, Division of Biological Drugs) but also include the Division of Traditional Chinese Medicines and Ethno-Medicines to reflect the importance of these in Chinese clinical practice (NPMA Department of Drug Registration 2019).

The International Council for Harmonisation of Technical Requirements for Pharmaceuticals for Human Use (ICH)

The ICH, based in Geneva, Switzerland was founded in 1990 by regulators and associated personnel from the USA, Europe and Japan. As the name implies, its objective is to harmonise procedures from different agencies with aim of making drug development more efficient through avoidance of duplication. Since it's founding, the number of members has expanded to include China and competent authorities worldwide. ICH recommendations are divided into Quality, Efficacy, Safety and Multidisciplinary Guidelines which provide detailed information for the key regulatory aspects of drug development in a readily accessible format (see ICH Guidelines 2025).

---

[8] See Pharmaceuticals and Medical Devices Agency (PDMA 2025)

International Coalition of Medicines Regulatory Authorities (ICMRA 2025)[9] and International Pharmaceutical Regulators Programme (IPRP 2025)[10]
The above organisations have been created to encourage the discussion of regulatory issues at a global level and to support the implementation of the ICH Guidelines, for example.

2.2.4 Pharmacoeconomics

Despite the obvious contribution that the biopharmaceutical industry makes to human health through its medicines and devices, it is also accused of profiteering at the expense of healthcare systems worldwide. While the last point is outside the scope of this technically orientated book, this section briefly covers some of the main issues in the discipline of pharmacoeconomics which applies a cost-effectiveness analysis (CEA) to drug pricing. It includes the costs involved in drug development, pricing models and the ways in which payment for very expensive medicines could save money in the long term. A detailed source of information on pharmacoeconomics can be found at the International Society for Pharmacoeconomics and Outcomes Research Inc. (ISPOR 2025).

2.2.4.1 Drug Development Costs

In 2016, the Tufts Center for the Study of Drug Development (TCSDD) published a report quoting a figure of $2.6Bn for the cost of developing a drug (DiMasi et al. 2016). As well as the actual development costs, the figure included the cost of failed programmes and the costs of tying up capital over many years before the drug can be marketed and begin to recoup the investment. Other cost estimates vary according to the published source and the therapeutic area being addressed, see for example Wouters et al. (2020) who come up with a median cost of $985 M. Each element of drug discovery has its own cost implications. The costs of clinical trials have been estimated in a number of studies; the one by Martin et al. (2017) for example, giving median costs of $3.4 M, $8.6 M and $21.4 M for phase I (with patients), phase II and phase III respectively. Cost implications for advanced medicinal products such as cell and gene therapies will be discussed later.

2.2.4.2 Customers

A patient requiring a marketed drug is rarely the direct customer. These will be the healthcare systems and managed care organizations that buy medicines in bulk and negotiate discounts from the manufacturers. The customer categories are listed below:

---

[9] See ICMRA
[10] See IPRP

Individual Patients
Physicians and Pharmacists
Government-Financed Healthcare Services
Private Healthcare Organizations

Social healthcare provision, private funding and insurance varies according to country. Britain's National Health Service (NHS) provides most of the country's healthcare from taxation and national insurance with a total expenditure of £181.7Bn in 2022–23 (The King's Fund 2023). Of this figure, £9.1Bn was spent on drugs, although provision was made for a COVID-19 inventory of £9.3Bn.

In the USA, Medicaid and Medicare Part D are administered by a government department, namely the Centers for Medicare and Medicaid Services (CMS). Medicaid is a social welfare system that provides healthcare support for those on low incomes, while Medicare is a social insurance system available to older people through both direct taxation and employee contributions. Total expenditure in 2022 was $944.3Bn and $805.7Bn for Medicare and Medicaid respectively. Private health insurance was $1289.8Bn the same year and total expenditure on drugs was $405.9Bn (CMS.gov NHE Factsheet 2022).

2.2.4.3 Pricing and Reimbursement

Healthcare markets are under great cost pressures, but the biopharma industry needs to be able to cover its costs and invest in new products. Public perception of high, or even excessive, drug prices may relate to the list price of a drug offered by manufacturers (wholesale acquisition price WAC) rather than a net price (average selling price ASP) that can be over 40% lower through negotiation; however, not all of the discounts will necessarily be passed on to patients (PharmExec.com 2020).

Reimbursement is the money that the holders of healthcare funding will pay the front-line suppliers of that care, the NHS in the UK, for example, or Medicaid/ Medicare in the USA. Funders will naturally try to keep prices as low as possible, often through using generic products that are substantially cheaper than patented medicines (see later). Organisations like the National Institute for Health and Care Excellence (NICE) in the UK have been established to objectively evaluate the cost and health outcomes of new medicines in terms of quality-adjusted life-years (QALYs). Regardless of discounting and given the increasing cost of medicines, particularly advanced therapies, there are some potential solutions to the tension between healthcare providers and manufacturers who need to recoup the costs of drug development and invest in future research. Some examples of this are given below:

Value-Based Arrangements

Value-based arrangements (VBAs) involve linking the pricing of drug treatments to clinical outcomes. This area is in an early stage of development but is particularly relevant for pricing and reimbursing expensive advanced therapies. In fact it is essential given that currently approved gene therapies are priced in millions of dollars. A detailed discussion of these issues and possible solutions has been published by Eichler et al. (2022). For example, risk-sharing pools are being considered where payers contribute a fixed amount and provide advanced therapies at a greatly discounted price.

Some therapies are potentially curative, so there is an argument to be made that a single expensive treatment (*one and done*) could be cost effective in the long run. One example is Betibeglogene autotemcel (Zynteglo) produced by Bluebird bio to treat Beta Thalassemia. The list price is $2.8 M, but if the gene therapy can stop the need for regular blood transfusions costing more than $6 M, it clearly would be cost-effective. Furthermore, Bluebird bio is committed to refund 80% of the treatment cost if patients do not become independent of transfusions (Fierce Pharma 2023).

Netflix Models

Netflix-models, based on the film and TV subscription service, are novel pricing and reimbursement agreements whereby manufacturers receive a fixed (recurring) payment for the supply of a new therapy, independent of the quantity sold (Hampson and Steuten 2024). A prominent example is the price for the hepatitis C antiviral drug sofosbuvir negotiated by the Australian government with the manufacturer Gilead. An upfront payment (subscription) of $776 M was made for five years' worth of the drug whose list price was $73,000. The resulting price per patient was between $7 and 12,000 thus representing a substantial saving (Moon and Erickson 2019). The model could also be applied to antibiotics where the economics of R&D are very different to mainstream drug development.

2.2.5 Intellectual Property (IP)

The main elements of IP are copyright, trade secrets, trademarks, bailments and patents. These are summarized below:

Copyright

This often relates to written material or images where the originator has an automatic right to ownership. From a pharmaceutical R&D perspective, this will most likely apply to material in technical or commercial publications.

Trade Secrets

These only last so long as they remain secret, i.e. are not disclosed to others without a prior legal contract, such as a confidentiality or non-disclosure agreement (CDA or NDA).

Trademarks

These can be registered in most countries and are used for company logos and related branding material.

Bailments
A legal term used to describe the property rights of a donor who gives physical materials to a third party. These materials may be biotechnology products, for example, in which there is a great deal of associated know-how that must be protected. A common bailment is the Material Transfer Agreement that occurs between laboratories in academia and industry.

Patents (or Letters Patent)
Legal instruments which are designed to prevent the exploitation of an invention by competitors The USA and UK system originated in eighteenth century England, in the reign of Queen Anne. Each country has its own patent law, but harmonization through the World Trade Organization encourages individual countries to comply with global regulations. The Agreement on Trade-Related Aspects of Intellectual Property Rights (TRIPS) has helped to bring the trade in pharmaceuticals under a global umbrella.

2.2.5.1 Patent Timelines

Patenting begins with an application (filing) to a national office (e.g. the US Patent & Trademark Office or the UK Intellectual Property Office). In the USA, it is possible to file a provisional patent that lasts for one year before making the decision to file a non-provisional patent (which is published after 18 months). This is a useful way of keeping the costs down while assembling the data required to proceed to a full patent, or to abandon the process at an early stage. As there is no such thing as an international patent, it is possible to use the Patent Cooperation Treaty (PCT) to file patents in multiple countries that are valid for about 30 months before having to be submitted to individual countries. Once a patent application is filed in a specified territory it has a lifespan of 20 years before the patent expires. This might seem like a long time but given the years that it takes to develop the drug the amount of patent protection left once it goes on the market could be significantly reduced. However, to compensate for the time taken to review and authorise a product, a maximum of 5 years extension of protection is possible in the USA and Europe.

There are other situations in which patent holders can gain beneficial extensions. For example, the FDA can grant *Pediatric Exclusivity*, i.e. an extra six months of patent protection if a drug can produce health benefits in children. Orphan Drugs are branded medicines developed for rare diseases. In the USA, the FDA provides seven years of market exclusivity for drugs that treat fewer than 200,000 people and the EMA, ten years of exclusivity for fewer than five in 10,000 people. Orphan drugs and rare diseases are covered later.

2.2.5.2 Composition of Matter Patents

These protect novel compounds that possess a unique chemical structure, have defined practical applications and be non-obvious to experts in the field. In practice, many different analogues of the compound must be laboriously documented along with demonstration of their utility, for example against a molecular target involved in a particular disease. Sometimes the patent will cover the formulation of a known compound that improves delivery.

Patenting of gene and cell-based therapies is a more complicated matter and is highly relevant to the advanced technologies described in this book. In the past, there was great deal of controversy about patenting DNA sequences since some biotech companies were considered to be patenting life against the spirit of scientific enquiry and holding up the development of new diagnostics and therapies. In 2013 a US court ruling in the case between the Association for Molecular Pathology and Myriad Genetics Inc. stated *that human genes cannot be patented because they are a product of nature.* This has not had an adverse effect on commercial activity since the human genome sequence has been in the public domain for many years. There is still some uncertainty however, about the patent position for individual products of biotechnology, but modified cells, nucleic acids etc. can certainly qualify for IP protection. A publication by Grassano et al. (2024) covers biotechnology patent categories in some detail. Some edited examples include:

- Compositions comprising non-specified tissues or cells, non-embryonic stem cells and genetically modified cells.
- Medicinal preparations containing genetic material which is inserted into cells of the living body to treat genetic diseases–Gene therapy
- Libraries contained in or displayed by micro-organisms or vectors
- Libraries containing nucleotides or polynucleotides, or derivatives thereof
- Libraries containing peptides or polypeptides, or derivatives thereof
- Methods of creating libraries, e.g. combinatorial synthesis: Biochemical methods, e.g. using enzymes or whole viable micro-organisms
- Bioinformatics, i.e. methods or systems for genetic or protein-related data processing in computational molecular biology (in silico methods of screening virtual chemical libraries; *in silico* or mathematical methods of creating virtual chemical libraries)

2.2.5.3 Process Patents

The chemical and biological processes that lead to patented small molecule and biotech drugs are often an important part of a company's patent estate. Sometimes companies will even attempt to patent manufacturing processes for out-of-patent drugs. Process patents are important for biotechnology; the humanization of monoclonal antibodies is a good example, where a highly competitive technology was developed in Cambridge, UK and patented by the UK's Medical Research Council technology transfer arm LifeArc thus creating a significant royalty stream.

2.2.5.4 Generic Competition

Once the patent on a branded prescription medicine expires, generic drug manufacturers can produce an identical product (in terms of the chemical composition of the active ingredient[11]) and then sell the product at a lower price. The 1984 Drug Price Competition and Patent Term Restoration Act (otherwise known as the Hatch-Waxman Act) in the USA allows generic manufacturers a shorter development time prior to approval compared with the originator of the branded product. The former are able to submit an Accelerated New Drug Application (ANDA) compared with the normal New Drug Application (NDA). The latter can benefit from patent extension to compensate for review times as mentioned in Sect. 2.2.5.1 earlier.

Biosimilars

Small molecule generic drugs are identical atom for atom with the branded equivalent, but this is not the case with biologicals due to the chemical nature of proteins, deamidation for example. Also, glycosylation differences can influence the pharmacokinetics of the drug, but it is not possible to generate these molecules in a totally uniform way using standard manufacturing processes. For this reason generic biologicals, or follow-on biologicals, are commonly known as *biosimilars*. The development of these generic proteins is inevitably more costly than with their small molecule equivalents, but the rewards are significant given the prominence of biologics in the list of top 10 best-selling drugs (see Chap. 1). Biosimilars are not just being made to copy large antibodies. Peptide mimetics of glucagon-like peptide-1 (GLP-1) liraglutide and semaglutide were approved for weight loss in 2014 and 2021 respectively. In China, the patent for liraglutide has already expired, and the one for semaglutide will do so in 2026. A liraglutide biosimilar is being developed in India for type 2 diabetes (Mallapaty 2024).

A useful guide to biosimilars can be found in Pfizer: See how biosimilars are developed (2021).

2.2.5.5 The Patent Cliff

There is an inevitable tension between developers of branded medicines and consumers of healthcare who want to save money, but whatever the pros and cons of the situation, generic competition has led to a dramatic loss of revenues for biopharma, otherwise known as the *Patent Cliff*. A forecast of losses from 2024 to 2028 has been published in IQVIA Global Use of Medicines Outlook to 2028 (2024) summarised in Table 2.5.

2.2.5.6 Life Cycle Management

Product life cycle management (PLM) can be employed in the following stages: product development, introduction, growth, maturity, and decline. The previous section on the Patent Cliff highlights the significant loss of revenues due to generic

[11] But see Biosimilars

Table 2.5 Projected revenue losses in $Bn for small molecule drugs (NCEs) and biologics; total $192Bn: $133.0Bn NCEs, $59.1Bn Biologics

Year	NCE	Biologic
2024	19.0	12.2
2025	21.6	14.1
2026	32.1	16.1
2027	25.7	7.5
2028	34.6	9.3

competition. It is therefore in a biopharmaceutical company's interest to manage the life cycle to maximize profits over the time available and to ensure that future programmes are in place to absorb the commercial shock of patent expiry.

Companies may undertake a process of *evergreening* through secondary patents to maximise the commercial returns on a branded medicine. These patents are used to extend the patent life of drugs through reformulation, finding new indications, or producing different chemical or physical forms. Manufacturing patents are also used as *patent thickets* to surround and protect the branded product. This approach has ensured that the monoclonal antibody Humira® for example has been able to resist competition from biosimilars for some years past the expiry of the original core patent in 2016. The originator AbbVie was granted 132 patents covering manufacturing, formulation and potential future formulations lasting until 2037. Producers of biosimilars did not pursue litigation but instead pay AbbVie royalties that will last until 2037 (Knox and Curfman 2022).

Second Medical Use Patents

When an approved drug is shown to work against a different disease to the one for which it was originally developed, it is possible to file a secondary medical use patent to protect the IP position. One well-known example is Pfizer's sildenafil (Viagra®) which was developed for angina but sold in a completely new market for treating erectile disfunction. The drug thalidomide gained notoriety through the birth defects it caused after being prescribed to pregnant women for morning sickness. However, it was shown later to be effective in treating multiple myeloma, so analogues such as lenalidomide have been developed for this purpose. More recently thalidomide has become a focus of interest due to its relevance to targeted protein degradation, something that will be covered in detail later in this book.

2.3 Current Pharmaceutical R&D

This section covers biopharmaceutical industry activity at the time of writing (2024–25), concentrating on the diseases being targeted and the medicines reaching marketing approval.

2.3.1 Therapeutic Areas

There are 17,000 diseases listed in the latest version of the WHO International Classification of Diseases 11th revision (ICD-1 2025). New conditions such as gaming addiction have been added since ICD-10, reflecting a very twenty-first century problem. The upheavals in education and normal life experienced by young people during the COVID-19 pandemic has left a legacy of mental conditions that only serves to highlight the already vital importance of drug discovery (and non-pharmacological therapies) for psychiatric medicine. This area has proven to be poorly served at present due to the difficulties in tuning drug responses to precise targets in the complex brain environment without causing undesirable side effects. Related to this is the problem of drug addiction, for example to opioids, which offers similar challenges for developers of future medicines. Of course, the major change in therapeutic focus was the COVID-19 pandemic, which diverted resources towards vaccines, small molecules and antibody therapies. It has also provided the impetus to develop messenger RNA (mRNA) technology to a point where it has been shown to be safe and effective, as well as lucrative for Pfizer and its BioNTec partner as well as Moderna.

There are clearly many different aspects of human biology and pathology that could be influenced by pharmaceutical products. Contraception, for example, which since the pill was introduced, has predominantly focussed on female biology. However, the recent identification of viability targets in sperm provide the opportunity for producing a male contraceptive e.g. (Ku et al. 2024).

Figure 2.2 summarises the main disease areas of interest to the biopharma industry.

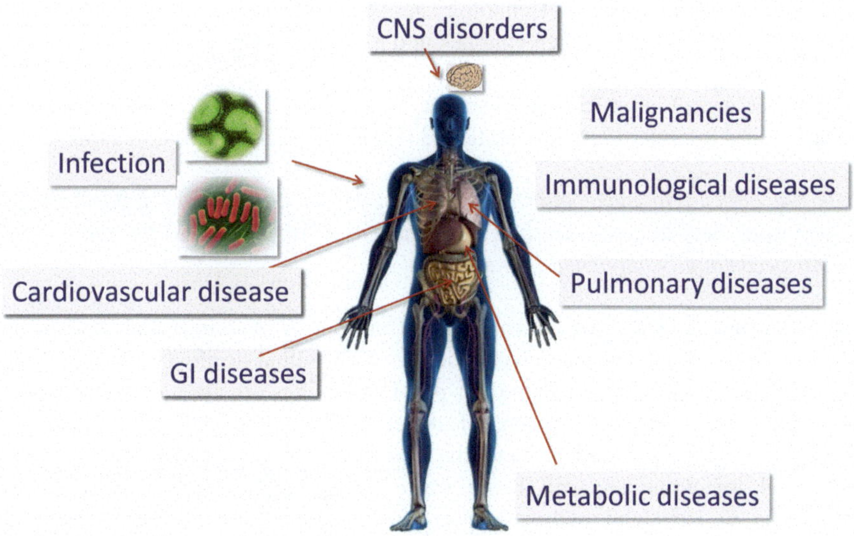

Fig. 2.2 Summary of major disease areas

This summary reflects the conditions affecting populations of all ages from birth (e.g. monogenic diseases) to older adults (e.g. cardiovascular, cancers and CNS disorders). Each category is covered in turn (starting at the top):

Central Nervous System (CNS) Disorders

Apart from the psychiatric conditions alluded to earlier, there are other CNS disorders that are currently poorly served with therapies but which provide future opportunities for drug development. Epilepsy and pain are two examples, along with degenerative conditions leading to dementia, i.e. Alzheimer's, Parkinson and Huntington's diseases. Advances in ion channel biology show promise for finding treatments for the former, while antibody and gene therapy may control the latter. However, neurodegenerative conditions are complex and difficult-to-treat, with many drug development failures; furthermore treatments may only be effective if a diagnosis is made before the disease has fully manifested itself.

Infection

Apart from more recent pandemic coronaviruses, there are plenty of viruses demanding attention, including HIV and influenza. In the former case, pre-exposure prophylaxis (PReP) with a cocktail of viral inhibitors prevents HIV infection and helps to relieve the burden of AIDS. This is a significant advance but falls short of the ideal of a vaccine. Trials of HIV vaccine candidates have been disappointing, with the problem of the virus infecting resting T-cells ready to be released later upon immune stimulation. There is a large body of research on immune responses to HIV, including the production of broadly neutralising antibodies that could be exploited in future therapies, so there are plenty of opportunities for developing new medicines for this area. Other viruses like influenza may be more amenable to vaccination strategies now that the mRNA technology has been validated, allowing for the facile introduction of multiple antigen epitopes into a single product.

Another aspect of infectious diseases is the rise of antimicrobial resistance (AMR). Sir Alexander Fleming mentioned this in his Nobel Lecture in 1945:

it is not difficult to make microbes resistant to penicillin in the laboratory by exposing them to concentrations not sufficient to kill them, and the same thing has occasionally happened in the body.

(quoted from the UK government report Confronting Antimicrobial Resistance 2024 to 2029 2025). The same report quotes 1.27 million deaths globally in 2019 as being due to AMR. Confronting this problem requires organisational changes in prescribing etc., the development of novel antibacterials and the commercial incentives required for biopharma to invest in costly and difficult R&D.

Fungal infections are still being treated with amphotericin B which was introduced in 1959 and with appropriate formulations, adverse events such as nephrotoxicity can be minimised. However, there is much drug development activity in this area, so new anti-mycotic approaches should be available in the future.

Cardiovascular Disease

This category covers cerebrovascular diseases and heart conditions (ischaemic and non-ischaemic). The projected figures for diseases shown in Chap. 1 indicate that cerebrovascular diseases are likely to increase significantly over the decades to follow, second only to malignancies.

Cerebrovascular disease is an umbrella term for stroke, or interruption of the brain's blood supply; this can be a transient ischemic attack (TIA) with temporary reduction in the supply of blood to the brain, or subarachnoid haemorrhage and vascular dementia resulting from insufficient blood supply to regions of the brain. Preventive measures include taking drugs to reduce hypertension and cholesterol, while treatment options usually involve surgery or *clot busting* drugs. Lifestyle changes are an obvious way of reducing risk, but more sophisticated approaches based on target identification, personalised medicine and new ways of controlling cholesterol levels may have an impact in the future. Some of these ideas will be discussed later in the book.

Gastrointestinal (GI) Diseases

In the last part of the twentieth century GI diseases featured largely due to the success of the anti-ulcerant drugs. Of course, excess production of stomach acid is one of many conditions affecting the alimentary tract from mouth to anus. Chronic inflammatory conditions such Crohn's disease and ulcerative colitis are of major interest to current biopharma, along with other conditions like pancreatitis, oesophagitis and irritable bowel syndrome (IBS). An increased understanding of the role of the microbiome in these diseases, plus target identification based on genetic studies has improved the outlook for delivering novel therapies.

Metabolic Diseases

Hypertension, elevated blood sugar, triglycerides and cholesterol and excess abdominal fat collectively fall under the umbrella term *metabolic syndrome* which can be a precursor to heart disease, stroke and type 2 diabetes. The last condition is widespread and of intense interest to biopharma. More recently attention has been focussed on improving weight-loss therapies, leading to major success with a family of appetite-suppressant peptide hormone mimetics. It is estimated that by 2035, more than half of adults worldwide will be obese, with a weight loss market forecast to be up to $100Bn by 2030 (Mullard 2024a). Nonalcoholic fatty liver disease (NAFLD) affects roughly 25% of US adults, 20% of whom can develop the more severe nonalcoholic steatohepatitis (NASH) that can lead to liver failure and possibly cancer. These conditions are related to type 2 diabetes and obesity, thus reinforcing the importance of these metabolic disease areas in drug development.

Pulmonary Diseases

Chronic obstructive pulmonary disease (COPD) affects an estimated 392 million people, three million of whom will die each year. The problem will only get worse as tobacco use decreases in affluent societies but increases in low-income countries, adding to their problems with air pollution. While preventative measures are obvious, their implementation is challenging, and the problem is not going to disappear readily. However, there are many opportunities for developing medicines both to treat the inflammatory condition itself and to help with smoking cessation. Other pulmonary diseases include asthma, bronchitis and pulmonary hypertension. Asthma has been a target for drug development for many decades, with products that give symptomatic relief or some control of inflammation. The goal of total disease modification maybe within reach, partly through the introduction of a new range of biologicals.[12]

Immunological Diseases

Some conditions with immune aetiology have been mentioned already, namely allergic asthma (a hypersensitivity disease) and Crohn's Disease (an autoimmunity). Chronic autoimmune conditions such as rheumatoid arthritis, multiple sclerosis and type 1 diabetes remain significant areas for biopharmaceutical development particularly since some biological agents, for example antibodies to tumour necrosis factor alpha (TNF-α), have shown disease modifying effects in rheumatoid arthritis and Crohn's. Inflammation is a common element of these and many other conditions which is why drug development in this area continues unabated.

Malignancies

Cancer is unquestionably the therapeutic area with the most R&D activity. The projection in Chap. 1 (Fig. 1.1) shows that malignant neoplasms will far exceed other conditions. An ageing population (see also Chap. 1) ensures that while cancer may remain controlled for most (but sadly not all) people in early life, the nature of the disease means that eventually a higher proportion of older people will be diagnosed with one form or another. UK cancer statistics (Cancer Research UK 2025) show more than 385,000 new cancer cases in the UK every year, i.e. more than 1000 every day (2017–2019). In 2018 almost half (45.5%) of all cancer cases were diagnosed at stage 3 and 4, which highlights the need for earlier diagnosis. By 2038–2040 there could be around 506,000 new cases of all cancers combined every year in the UK alone. Improvements in treatment outcome vary according to the type of cancer as illustrated in Fig. 2.3 using data from 1971–2 to 2010–11. Lung cancer presents a dire scenario and pancreatic cancer has been effectively unchanged over decades. However, big improvements have been made with testicular, prostate, breast, melanomas and blood cancers. This can be seen as a positive outcome, but one gained at the expense of an often-gruelling treatment regimen involving surgery, radiotherapy and cytotoxic chemicals. Oncology will feature throughout the rest of this book in illustrating how technologies for discovering novel drug targets and medicines are being applied through biopharmaceutical R&D.

[12] The basic science behind these is not new, as it was a key part of my own research into asthma and allergies at what was then Glaxo Group Research

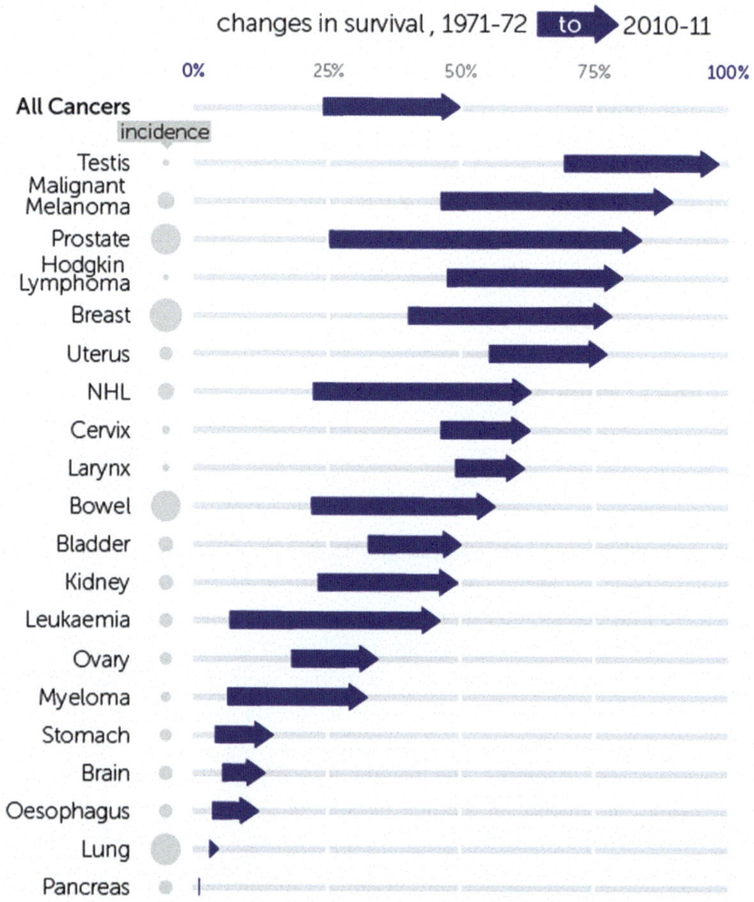

Fig. 2.3 UK survival statistics for a range of cancer types. (Figure courtesy of Cancer Research UK)

Orphan Diseases

Rare diseases are rare, but rare disease patients are numerous (Orphanet 2025). There are thousands of rare diseases, with the term *orphan disease* often used as a synonym. The FDA defines a rare disease as one affecting fewer than 200,000 individuals, and for the EMA, fewer than 5 in 10,000. The 1983 Orphan Drug Act in the USA allowed financial incentives for drug developers as compensation for a small market size. These incentives include R&D tax credits, waived FDA user fees and the potential for a seven-year period of marketing exclusivity if the drug is approved for use in the rare disease or condition. Miller et al. (2022) surveyed FDA-designated orphan drugs representing 3269 unique products listed between 1983 and 2019. Of these, 508 had an orphan-designated approval and were derived from drug repositioning strategies or parallel indication development programs for recently developed drugs. The main therapeutic areas addressed by approved drugs were, in decreasing order of size, oncology, neurology, infectious diseases, metabolic disorders and haematology.

High profile, but still rare, conditions such as cystic fibrosis, Duchenne muscular dystrophy, and sickle cell disease are paediatric onset diseases associated with single gene defects. Orphan drug designations for these conditions represented 27% of the total in the 2010s. With increased genetic information about these monogenic diseases and the molecular tools being developed to correct them, the whole area will certainly expand to the benefit of patients but also raise hard financial questions for healthcare providers.

2.3.2 Clinical Pipelines in 2023–24

A breakdown for the end of 2023 of drug development pipelines from preclinical to approval is shown in Table 2.6[13] providing a snapshot of biopharma interest within different therapeutic areas.

Oncology is the largest indication with 350 approved products. Improvements in treatment have been made over the years, but not for all cancers (see Fig. 2.5). The challenge is to find cures rather than just extending disease free progression.

Metabolic disease has become a high-profile area due to the success of GLP-1 agonists like semaglutide (Ozempic) in promoting weight loss. Where success has become apparent over a relatively short time frame (as with mRNA vaccines, CAR T-cells etc) there is an incentive to put extra resources into overcoming technical and other difficulties. For example, the weight loss drugs also promote undesirable muscle wasting. In preclinical studies by BioAge, a combination of Eli Lilly's GLP-1 agonist Mounjaro and the apelin receptor agonist azelaprag (BioAge) simultaneously boosted weight loss and preserved lean muscle. As a result, phase 2 studies are underway with this combination.

2.3.3 Current Drug Approvals

Some of the drug candidates approved by regulators at a particular time will become best-selling medicines in due course. Others will of course fall by the wayside. In any event, the types of product and their therapeutic indications give a feel for how biopharma R&D might develop over the next few decades, and which is presented in more detail in Chap. 6. Some examples of FDA approvals in 2023 are presented in Tables 2.7 and 2.8.

[13] According to a report Global Trends in R&D 2024 by IQVIA, 28% of trial starts in 2023 (out of 4873) were from Chinese headquartered companies

Table 2.6 Drug development pipeline for major therapeutic areas end 2023

	Preclinical	Phase I	Phase II	Phase III	NDA/BLA	Approved	Suspended
Oncology	424	417	688	292	67	350	66
Respiratory	20	7	33	15	8	20	6
Psychiatry	28	31	46	36	9	28	7
Neurology	104	60	126	93	12	109	14
Metabolic	74	22	79	58	33	56	13
Infectious disease	89	59	91	74	32	78	22
Haematology	37	10	23	16	17	13	1
Endocrine	45	25	74	45	23	25	18
Cardiovascular	38	18	31	39	15	24	3
Autoimmune/immunology	176	41	115	83	33	92	16

Data taken from Hodgson (2024) with permission. *NDA* new drug application, *BLA* biologic licence application

Table 2.7 CDER approvals in 2023 taken from Mullard (2024b) with permission

Drug action	Indication
Monoclonal antibodies	
Amyloid-ß-targeted mAb	Alzheimer's disease
C5-targeted mAb	CHAPLE disease
FcRn-targeted mAb	AChR- or MUSK-antibody-positivie gMG
IL-17A/F-targeted mAb	Plaque psoriasis
IL-23-targeted mAb	Ulcerative colitis
PD1-targeted mAb	Merkel cell carcinoma
PD1-targeted mAb	Nasopharyngeal carcinoma
RSV F-protein-targeted mAb	RSV lower respiratory tract disease
Bispecific antibodies	
BCMAxCD3 T-cell engager	Multiple
CD20xCD3 T cell engager	DLBCL and high grade B-cell lymphoma
CD20xCD3 T cell engager	DLBCL and high grade B-cell lymphoma
GPRC5DxCD3 T-cell engager	Multiple
Recombinant proteins	
Growth hormone analogue	Growth hormone deficiency
PEGylated recombinant α-galactosidase	Fabry disease
Recombinant α-glucosidase	Pompe disease
Recombinant α-mannosidase	Non-CNS a-mannosidosis
Recombinant leukocyte growth factor	Neutropenia
RNA-based drugs	
LDHA-targeted siRNA	Primary hyperoxaluria type 1
SOD1-targeted ASO	Amyotrophic lateral sclerosis
TTR-targeted ASO	hATTR with polyneuropathy
C5-targeted aptamer	Geographic atrohpy secondary to AMD
Small molecules	
5HT1A receptor agonist	Major depressive disorder
AKT kinase inhibitor	Breast cancer
Antifungal	Candidemia and invasive candidiasis
BTK inhibitor	Mantle cell lymphoma
CGRP receptor antagonist	Migraine
Complement C5 inhibitor	AChR-antibody positive gMG
Complement factor B inhibitor	Paroxysmal nocturnal haermoglobinuria
Corticosteroid	Duchenne muscular dystrophy
CXCR4 inhibitor	Haematopoietic stem cell mobilisation
Ectoparasiticide	Demodex blepharitis
Endothelin and angiotensin II receptor antagonist	Primary IgA nephropathy
ER antagonist	ER-positive, HER2-negative ESR-1 mutant breast cancer
FLT3 kinase inhibitor	AML
GABAA receptor PAM	Postpartum
Gamma-secretase inhibitor	Desmoid tumours

(continued)

Table 2.7 (continued)

Drug action	Indication
HIF-PH inhibitor	Anaemia
JAK1/2, ALK2 inhibitor	Myelofibrosis in adults with anaemia
JAK3 inhibitor	Alopecia areata
Mechanism unknown	Epidermolysis bullosa
Neurokinin 3 receptor antagonist	Menopause flushes
NRF2 activator	Friedrich's ataxia
PI3Kδ inhibitor	Activated PI3Kδ syndrome
Radioactive diagnostic agent	PET imaging prostate cancer
Retinoic acid receptor agonist	Fibrodysplasia ossificans progressiva
ROS1 and TRK1 kinase inhibitor	ROS1-positive NSCLC
S1P receptor modulator	Ulcerative colitis
SARS-CoV-2 main protease inhibitor plus CYP3A inhibitor	COVID-19
Semifluorinated alkane	Dry eye disease
SGLT1/2 inhibitor	Heart failure
SGLT2 inhibitor	Type 2 diabetes
ß-lactam + β-lactamase	Antibiotic plus β-lactamase inhibitor
Thiadiazinane antimicrobial plus anticoagulant	Catheter-related bloodstream infections
Unknown mechanism	Rett syndrome
VEGFR1/2/3 kinase inhibitor	Colorectal cancer

CDER Center for drug evaluation and research

Out of the 71 CDER and CBER-approved products, 34 were small molecules,12 antibodies, 8 recombinant proteins, 5 gene therapies, 5 vaccines, 4 RNA-based drugs and 3 cell therapies. Four of the antibody candidates are bispecific mAbs, a recently introduced modality that engages both tumour antigens and the CD3 T-cell surface molecule. This has the effect of bringing cytotoxic T cells into proximity with the tumour to enhance the efficiency of cell killing. Cell-based and gene therapies include a transplanted microbial preparation for *C.difficile* infections and the first approved CRISPR-Cas9 gene editing therapy. Two of the RNA-based drugs are based on antisense oligonucleotides (ASOs), one on small interfering RNA (siRNA) and one aptamer that works by binding to a protein target. While it is encouraging to see these agents approved, particularly after a long and chequered history, the markets (possibly excepting macular degeneration) are still quite limited.

2.4 The Drug Discovery Pipeline

This chapter concludes with a description of the drug discovery pipeline i.e. the key components of the discovery and development process. A summary is shown in Fig. 2.4.

This simple arrangement is designed for clarity and brevity, but it represents a highly simplistic view of drug discovery and development that in reality is a complex set of processes and interrelationships that do not necessarily occur within

Table 2.8 CBER approvals in 2023 taken from Mullard (2024b) with permission

Drug action	Indication
Cell therapy	
Allogeneic pancreatic islets	Type 1 diabetes patients unable to approach target HbA1c
Nicotinamide modified allogeneic haematopoietic progeniture cell therapy	Neutrophil recovery in patients with haematologic malignancies
Seres bacterial spore suspension in capsules for oral administration	*C. difficile* recurrence
Gene therapy	
AAV-based gene therapy with a micro-dystrophin	DMD patients aged 4–5
AAV-based gene therapy with modified factor VIII transgene	Haemophilia A
CRISPR–Cas9-based BCL11a-gene-editing therapy	Sickle cell disease
HSV-based gene therapy with COL7A1 transgene	Skin wounds in COL7A1-mutated DEB
Lentivirus-based gene therapy with HbAT87Q transgene	Sickle cell disease
Vaccines	
Anthrax vaccine adsorbed, adjuvanted	Protective antigen protein-based vaccine post-exposure prophylaxis
Bivalent stabilized prefusion F subunit vaccine	RSV prevention, in over 60 s and pregnant women
Conjugate and recombinant protein vaccine pentivalent	Meningococcal disease prevention
Live, attenuated chikungunya virus vaccine	Chikungunya virus prevention
Stabilized prefusion F subunit vaccine	RSV prevention, in over 60 s
Recombinant proteins	
Prothrombin complex concentrate containing factors II,VII,IX X proteins C and S	Urgent warfarin reversal
rADAMTS13	cTTP
Recombinant antihaemophilic factor	Haemophilia A

CBER Center for biologics evaluation and research

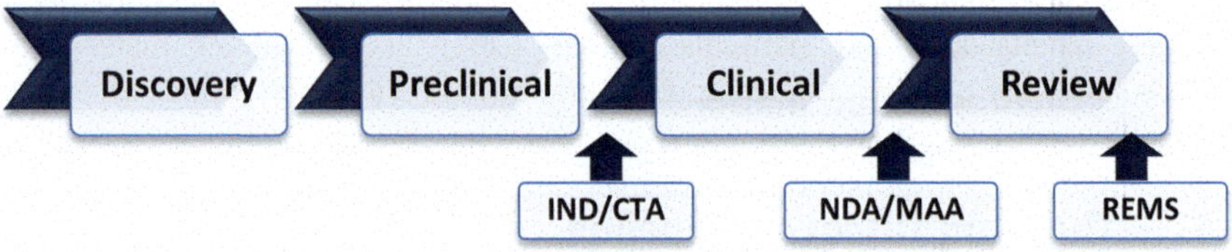

Fig. 2.4 Schematic diagram of drug discovery pipeline. *IND/CTA* investigational new drug/ clinical trial authorisation, *NDA/MAA* new drug application/marketing authorisation application, *REMS* risk evaluation and mitigation strategy

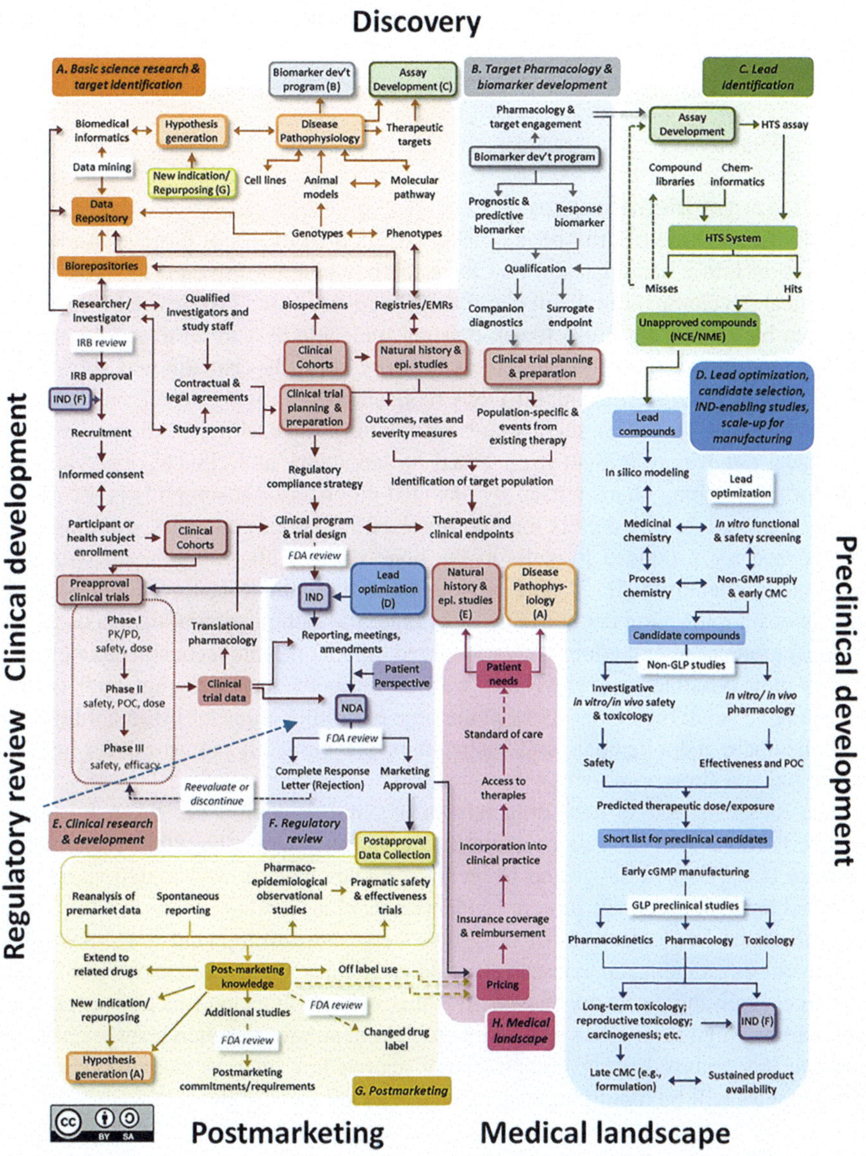

Fig. 2.5 *Drug Discovery, Development and Deployment Map* for small molecules. (Reproduced from Wagner et al. 2018 with permission. Boxes: Discovery orange, grey, green; Preclinical development, blue; Clinical development, pink; Regulatory review, mauve; Postmarketing, yellow; Medical landscape, dark pink)

a linear timescale. Wagner et al. (2018) have presented a more realistic version consisting of an accurate and customizable *Drug Discovery, Development and Deployment Map* (4DM), one of which is illustrated in Fig. 2.5. This resource, along with other information from regulators' websites and other references provides a great deal of detail. However, this chapter just summarises the key elements of the drug discovery pipeline, and in doing so, sets the scene for the rest of this book.

2.4.1 Discovery

2.4.1.1 Target Identification

The discovery phase of drug development covers the process of turning an idea or observation into a candidates(s) that have the potential to move all the way from preclinical development to the marketplace. Discovery could take several forms, for example by identifying a use for an existing molecule by careful observation and some lateral thinking. In this way, the laboratory curiosity cisplatin was shown by Rosenburg et al. (1965) to inhibit DNA replication in growing *E.coli* and led to a new generation of cancer therapeutics (Rosenburg et al. 1965). Alternatively, a biological target may present itself based on laboratory and clinical observations. The form of asthma characterised by elevated numbers of eosinophils is a case in point. These white blood cells may be involved in the inflammatory pathology of asthma and are produced by cells in the bone marrow after stimulation with the cytokine interleukin-5 (IL-5). The obvious point that inhibiting IL-5 action should reduce eosinophil numbers and may ameliorate the asthma symptoms encouraged the development of antagonist drugs; it turned out that despite a considerable effort, it was only possible to achieve this with an antibody approach, not with orally bioavailable small molecules. Now at the time of writing, three anti IL-5 antibodies, benralizumab, mepolizumab and reslizumab, are clinically approved for severe eosinophilic asthma.

Genetics can lead to new drug targets or confirm approaches taken by other routes. For example, the role of amyloid beta in Early Onset Familial Alzheimer's Disease (EOAD) was suggested by genetic studies that showed mutations in the amyloid beta A4 precursor protein (APP) (Goate et al. 1991).

One of the key objectives of drug discovery and development is to make the process more efficient and to create effective medicines in the shortest period of time possible. From the discovery perspective, this will involve target identification and validation at high throughput using tools such as genomics to generate large datasets that can be analysed using computational approaches such as machine learning. These topics will be highlighted later in the relevant parts of the book.

2.4.1.2 Choosing the Intervention

The example of IL-5 antagonists has already been mentioned in the context of choosing which drug type to use in a particular situation. The cytokine-receptor interaction (i.e. protein-protein interaction) has proven to be too difficult to disrupt with a small molecule, so a monoclonal antibody (mAb) was developed based on many years' experience gained in manufacturing and using these agents. The data presented so far about top selling medicines and recent FDA approvals highlight the fact that, while most are small molecules, there are a significant number of mAbs in the pipeline. Alternatively, a disease may occur because of a genetic defect, in which case the most viable approach could involve gene therapy. Lastly, cell therapy may provide the best option, for example by engineering stem cells to repopulate the haematopoietic system, establish beneficial organisms in the microbiota, or to boost the immune system in cancer. One thing that is clear at the end of the first quarter of the twenty-first century: there are no limits to the type of intervention being developed by the biopharma industry in contrast to the almost exclusive focus on small molecules in earlier years.

2.4.2 Preclinical Development

The phase between drug candidate selection for further development and filing an IND or CTA application for first time administration to humans is fraught with difficulties. There are a number of different stages where the drug could fail to meet the criteria of safety etc., so this is sometimes called *the valley of death*. These obstacles are there to be overcome however, so new technologies, including organoid cultures and machine learning, may improve the situation and these are covered in Chap. 7. Meanwhile, the basic processes of preclinical development are outlined in the following sections.

2.4.2.1 Manufacturing

The physical production of the drug candidate must be cost-effective, otherwise the *cost of goods* i.e. the actual cost of the material will be too high to be commercially viable. Issues of cost are particularly acute with advanced therapy medicinal products (ATMPs) given the complexity of manufacture and quality control. However, small molecule production has its own issues, if a large number of steps are required for their synthesis, or key components such as catalysts are expensive. For example, *every pharmaceutical we produce at some point or another has a palladium-catalysed step in it* (Per-Ola Norrby quoted in Remmel (2022). Palladium is expensive and toxic but indispensable for creating carbon-carbon bonds. The problem is highlighted by the example of producing 3 kilograms of a drug molecule where the ingredients cost $250,000, the palladium catalyst alone $100,000 and purifying it out of the product another $30,000 (Fuchß in Remmel 2022). A challenge for the future is to find alternative catalysts or to completely remove the palladium contaminating the manufactured products, something that is yet to be achieved.

ATMP manufacture is highly complex as exemplified by Pfizer/BioNTec's mRNA vaccine against COVID-19 (reviewed by Warne et al. 2023). The product required approximately 280 unique components for what is basically *in vitro* transcription off a plasmid template using modified nucleotides and encapsulating the resulting mRNA in a lipid coat.

2.4.2.2 Formulation

Formulation studies begin at an early stage, although the final dosage form may not be developed until later in the clinical trial phase. The objective is to ensure that the

Table 2.9 Solubility and permeability data for small molecule drugs cited by Bhalani et al. (2022)

BCS class	% Marketed drugs	% Pipeline drugs
I- high solubility, high permeability	35	5–10
II- low solubility, high permeability	30	60–70
III- high solubility, low permeability	25	5–10
IV- low solubility, low permeability	10	10–20

active pharmaceutical ingredient (API) is made available to the body in a consistent way. This involves combining the API with inert excipients to form tablets, capsules, inhaled preparations or liquids according to the type of molecule and therapeutic target. One of the major issues to be overcome is the solubility of the API in water. For oral absorption, the compound should ideally have high water solubility and intestinal permeability, factors considered in the Biopharmaceutics Classification System (BCS). Bhalani et al. (2022) have listed these properties for marketed and pipeline drug candidates which are shown in Table 2.9.

The table highlights the problem of poor compound solubility, something that can terminate further development. Of course, there are various approaches being used to overcome this problem, including nanotechnology.

Most biological drugs, including gene therapy vectors, are administered by injection of a ready-to-use liquid formulation; however, many are unstable in an aqueous environment requiring either freezing or drying, or the addition of excipients as stabilisers. Formulations for cell therapies have their own issues, including the need for cryopreservation.

2.4.2.3 Drug Metabolism and Pharmacokinetics (DMPK)

Pharmacokinetics and Pharmacodynamics

Pharmacokinetics (PK) measures the effect of the body on the drug and pharmacodynamics (PD), the effect of the drug on the body, summarised in Fig. 2.6. Both of course are critical parameters and can influence the outcome of a preclinical development programme. Sometimes the acronym ADME is used from pharmacokinetics as it describes:

Absorption: entering the body
Distribution: moving through the body
Metabolism: changing within the body
Excretion: leaving the body

Details about how PK studies are carried out, including measured parameters such as *area under the curve* (AUC) and the animal species used are covered elsewhere. A workflow is briefly summarised in Kovalainen (2020).

Drug Metabolism

The liver is a primary site of drug metabolism, a series of enzymatic reactions designed to break down ingested compounds into molecules that can be excreted

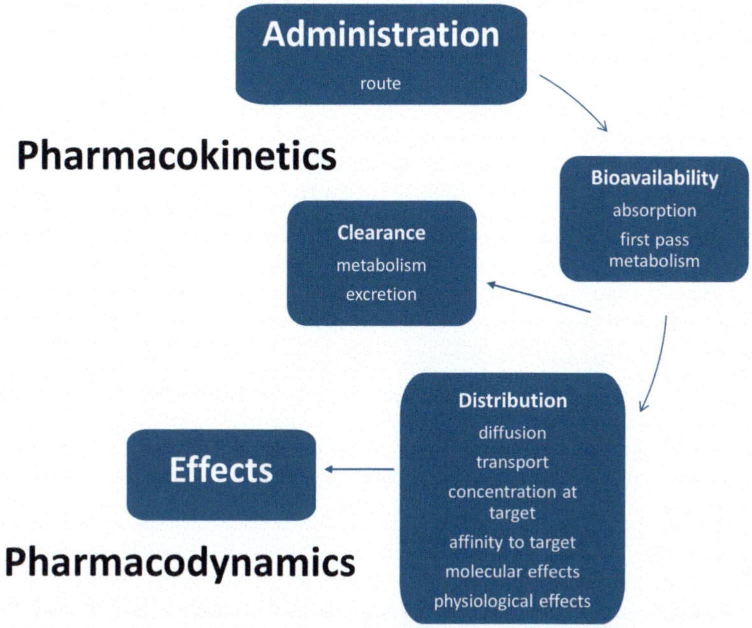

Fig. 2.6 Key elements of Pharmacokinetics and Pharmacodynamics. (Modified from Doogue and Polasek 2013 with permission)

in the urine or faeces. The main enzymes involved are the cytochrome P450 metalloenzymes whose variety is genetically determined and has an impact on drug interactions and safety (see later). If an orally administered drug is broken down too rapidly in the liver, i.e. has a high *hepatic clearance*, it will have to be delivered by a parenteral route. Alternatively, the medicine could be in the form of a *prodrug*, one that only becomes active once modified in cells or organs such as the liver. Finally, drug metabolites produced in the liver (or elsewhere) may be toxic, something that has to be investigated using cultured cells (see Chap. 7).

The nucleoside analogue remdesivir used to inhibit the RNA dependent RNA polymerase in the SARS-Cov-2 virus is an example of a prodrug (metabolised by esterases, amidases and kinases in cells) to produce the active molecule GS-441524. It also undergoes rapid hepatic clearance, so has to be administered by injection.

Zhang and Tang (2018) provide a useful overview of drug metabolism, including measures taken in medicinal chemistry to overcome problems. For example, metabolic switching can be used to remove metabolically labile features in the drug molecule by substituting with more stable functional groups. Molecules may also be designed to interfere with the active site of a particular CYP450 molecule. Of interest is the introduction of deuterated drugs in which hydrogens are substituted with the deuterium isotope forming a stronger bond (see also Chap. 6). There is a growing literature on deuterating existing drugs or developing new molecules from scratch to improve pharmacokinetics. Remdesivir, with its high lability in the liver, has been deuterated to form VVV116, an orally active molecule currently undergoing clinical trials (Di Martino et al. 2023).

2.4.2.4 Safety Pharmacology

Safety pharmacology is essentially a regulatory requirement for ensuring that a drug candidate is likely to be safe when administered to humans.

The following quote from Pugsley et al. (2008) states the issue succinctly: *the possibility that a drug may be of benefit informs the regulatory process to take an optimistic stance. In safety, the possibility (from preclinical studies of dubious validity) that a drug may be unsafe, informs the regulatory process to take a pessimistic stance. Although the latter might mean the delay or preclusion of useful drugs entering clinical use, it might also mean the prevention of harmful drugs doing likewise. The task for scientific Safety Pharmacology is to provide better evidence to direct the decision making.*

Small molecule drugs are administered to selected animal species to observe the effects on key physiological functions. The procedures are listed by the ICH as a *core battery* that includes a screen to account for the potentially life-threatening cardiac syndrome *torsades de pointes* (TdP). The fact that a number of marketed drugs have this liability accelerated the use of safety pharmacology in the 1990s.

The ICH core battery (taken from S7A guideline) is summarised in Table 2.10.

The guidelines include a series of follow-up and supplementary studies to cover different organ systems, known issues with similar compounds etc.

2.4.2.5 Toxicology and Secondary Pharmacology

Toxicology examines the effects of a drug candidate on body structures and includes carcinogenicity, genetic toxicology, reproductive toxicology and immuno-toxicology. Various protocols are described in the ICH Guidelines (S1B(R1), S2(R1), S4, S5(R3) and S8.

Secondary pharmacology involves testing drug candidates against off-target molecules such as receptors and enzymes. Drugs are normally evaluated in this way at an early stage to get some idea of issues that could arise later in development. Recent advances in thinking about these, and other issues in preclinical development

Table 2.10 Core battery of safety pharmacology studies (ICH-S7A)

System	Investigations
Central nervous system	Motor activity, behavioural changes, coordination, sensory/motor reflex responses, body temp
	Functional observation battery (FOB) modified Irwin's or other appropriate test
Cardiovascular system	Blood pressure, heart rate, and electrocardiogram
	Ex vivo evaluations, including methods for repolarization and conductance abnormalities
Respiratory system	Respiratory rate and (e.g., tidal volume or haemoglobin oxygen saturation

will be covered in Chap. 7. One of the most contentious relates to the use of animals in preclinical development. Recent FDA guidance allows developers to avoid using animals if it can be reasonably be shown that safety is unlikely to be compromised. The pros and cons are discussed in Wadman (2023) and see also Chap. 7.

2.4.2.6 Drug Interactions

The CYP450 family of enzymes are fundamentally important for drug metabolism and can also be involved in drug interactions, i.e. the possible effects of combining more than one medicine (or medicine and a foodstuff). One agent may inhibit a CYP450 enzyme thus allowing a second drug to increase in concentration. Alternatively, CYP450s can be induced in the liver, thereby decreasing the concentration of a second agent. These effects could obviously be deleterious, but also an advantage that can be exploited. For example, nirmatrelvir an inhibitor of the SARS-CoV-2 main protease, is administered in combination with the antiviral ritonavir. The latter is inactive against the coronavirus but acts as a CYP3A inhibitor, thus significantly increasing the plasma concentration of the active drug. Detailed information about drug interactions, including the role of metabolic enzymes other than CYP450s, is found in ICH guidelines M12.

2.4.2.7 The IND/CTA Application

The Investigational New Drug Application (IND) for the FDA, or Clinical Trial Authorisation (CTA) Application for the EMA are based on all the preclinical data accumulated in the drug development pipeline thus far. A key requirement is the dose estimate for first time in human (FTIH) trials. As might be expected, this is not always straightforward and there are a number of possible approaches. Shen et al. (2019) detail the requirements, including the different methods used to estimate the FTIH dose summarised as follows:

Maximum recommended safe starting dose (MRSD)
Similar mechanism of action to known drugs (MOA)
Minimum anticipated biologic effect level (MABEL)
Pharmacokinetic model
Pharmacokinetic/pharmacodynamic model

2.4.3 Clinical Development

According to the clinical trials database clinicaltrials.gov, on 11th July 2024 there were a total of 384,277 interventional trials of which 194,461 were for assessing small molecules or biologicals. The clinical development stage will reveal whether the drug candidate has activity against the chosen target and ideally improve patient outcomes by treating the disease of interest. Full details about the design and conduct of clinical trials, including biostatistics, can be found elsewhere, but the basic phases are summarised in Table 2.11.

Some developments in clinical trial design and implementation will be covered in Chap. 7. At this point it is worth mentioning the potential benefit of running phase

Table 2.11 Clinical trial phases

Clinical phase	Description	Timeframe
Phase 0	Microdosing	Weeks
Phase I	Human pharmacology	Weeks
Phase II	Exploratory	Months
Phase III	Confirmatory	Years
Phase IV	Post marketing studies	Years
Phase V	Post marketing surveillance	Years

0 trials to gain confidence in exposing human subjects to candidate drugs and thus avoid the use of animal models. Phase 0 studies, or exploratory clinical trials, evaluate subtherapeutic exposures of new drugs in FTIH studies. Originally established in 2003 to evaluate pharmacokinetics, they can also explore the pharmacodynamics and mechanisms of action (MOA) of drug candidates. Phase 0 trials use human subjects instead of animals to bridge the preclinical stage and early proof of concept (i.e. Phase Ib or Phase II). Burt et al. (2020) have reviewed the current status in detail and ICH guidelines are to be found in ICH M3(R2).

2.4.3.1 Pharmacovigilance

According to the WHO:

Pharmacovigilance is the science and activities relating to the detection, assessment, understanding and prevention of adverse effects or any other medicine/vaccine related problem.

Pharmacovigilance can be passive or active. Passive surveillance relies upon unsolicited communication by healthcare professionals or consumers to a company, regulatory authority or other organisation that describes one or more adverse drug reactions in a patient who was given one or more medicinal products. They play a major role in alerting the sponsor to rare adverse events that were not detected in earlier clinical trials or other pre-marketing studies. Active surveillance seeks to ascertain the number of adverse events via a continuous pre-organised process. This might involve follow-up of patients treated with a particular drug through a risk management program, gathering data from selected clinical (sentinel) sites, or questionnaires and disease registries. As with other topics in this chapter, more details are available elsewhere, including ICH E2E Pharmacovigilance Planning (Pvp) 2025.

2.4.4 Regulatory Approval and Marketing

2.4.4.1 Regulated Procedures

Good Laboratory Practice (GLP)

In the mid-1970s, it became clear to the FDA that laboratory procedures were being undertaken in an inconsistent way with possible negative consequences. As a result, a legally binding addition to the Code of Federal Regulations was introduced in 2005. It requires attention to:

Resources: organization, personnel, facilities, and equipment
Characterization of test items and test systems
Protocols and standard operating procedures (SOPs)
Raw data, final reports, and archives
Quality Assurance: independent monitoring of research processes

Good Clinical Practice (GCP)
From the introduction to the ICH Guidelines E6(R2):

Good Clinical Practice (GCP) is an international ethical and scientific quality standard for designing, conducting, recording and reporting trials that involve the participation of human subjects. Compliance with this standard provides public assurance that the rights, safety and well-being of trial subjects are protected, consistent with the principles that have their origin in the Declaration of Helsinki, and that the clinical trial data are credible.

Good Manufacturing Practice (GMP)
This is also termed *current Good Manufacturing Practice* (cGMP) to highlight the fact that older procedures and equipment may need to be updated. The ICH Q7 guideline covers GMP regulations for small molecules and biological products, but not cell or gene therapies. The EMA's regulations on Advanced Therapy Medicinal Products (ATMPs) are found in EMA (2019) The FDA regulates the use of tissues as therapeutics through the Good Tissue practices (GTP) regulations (FDA (2024).

2.4.4.2 Marketing Authorisation

Once the drug candidate has shown efficacy and safety in two pivotal phase III clinical trials, a marketing application can be made to the FDA (NDA, New Drug Application) or EMA (MAA, Marketing Authorisation Application). The enormous amount of supporting information can be incorporated into the Common Technical Document (CTD), see ICH M4: *The Common Technical Document* 2025. Again, more details can be found about the approval process itself, including priority review, accelerated approval etc. Of more recent interest, regulators have created rapid authorisation procedures to deal with serious infectious diseases, bioterrorism and diseases requiring urgent attention. The FDA has Emergency Use Authorisations (EUAs) which were notably used during the COVID-19 pandemic for vaccines and antivirals. The EMA equivalent is the Conditional Marketing Authorisation (CMA), which like the EUA, is valid for 1 year and addresses life-threatening diseases and public health emergencies. The applications need less comprehensive clinical data than normally required, where the benefit of immediate availability of the medicine outweighs the risk inherent in the fact that additional data will still be required.

References

Bhalani DV et al (2022) Bioavailability enhancement techniques for poorly aqueous soluble drugs and therapeutics. Biomedicines 10:2055. https://doi.org/10.3390/biomedicines10092055. Accessed 19 Mar 2025

BR Business Research Insights (2024). https://www.businessresearchinsights.com/market-reports/pharmaceutical-aseptic-fill-finish-cmo-market-100066. Accessed 19 Mar 2025

Burt T et al (2020) Phase 0/microdosing approaches: time for mainstream application in drug development? Nat Rev Drug Disc 9:801–818. https://doi.org/10.1038/s41573-020-0080-x. Accessed 19 Mar 2025

Cancer Research UK (2025). https://www.cancerresearchuk.org/health-professional/cancer-statistics/incidence#heading-Zero. Accessed 19 Mar 2025

CMS.gov NHE Factsheet (2022). https://www.cms.gov/data-research/statistics-trends-and-reports/national-health-expenditure-data/nhe-fact-sheet. Accessed 19 Mar 2025

Confronting Antimicrobial Resistance 2024 to 2029 (2025). https://www.gov.uk/government/publications/uk-5-year-action-plan-for-antimicrobial-resistance-2024-to-2029/confronting-antimicrobial-resistance-2024-to-2029. Accessed 19 Mar 2025

Corporate Finance Institute (CFI) (2025). https://corporatefinanceinstitute.com/resources/valuation/how-vcs-look-at-startups-and-founders/. Accessed 19 Mar 2025

Di Martino RMC et al (2023) Deuterium in drug discovery: progress, opportunities and challenges. Nat Rev Drug Disc 22:539–561. https://doi.org/10.1038/s41573-023-00703-8. Accessed 19 Mar 2025

DiMasi JA et al (2016) Innovation in the pharmaceutical industry: new estimates of R&D costs. J Health Econ 47:20–33. https://doi.org/10.1016/j.jhealeco.2016.01.012. s

Doogue MP, Polasek TM (2013) The ABCD of clinical pharmacokinetics. Ther Adv Drug Saf 4:5–7. https://doi.org/10.1177/2042098612469335. Accessed 19 Mar 2025

Eichler H-G et al (2022) Precision reimbursement for precision medicine: using real-world evidence to evolve from trial-and-project to track-and-pay to learn-and-predict. Clin Pharmacol Ther 111: 52–62. https://doi.org/10.1002/cpt.2471

EMA (2019) Guideline on quality, non-clinical and clinical requirements for investigational advanced therapy medicinal products in clinical trials. https://www.ema.europa.eu/en/documents/scientific-guideline/draft-guideline-quality-non-clinical-clinical-requirements-investigational-advanced-therapy-medicinal-products-clinical-trials-first-version_en.pdf. Accessed 19 Mar 2025

EMA Regulatory Science Strategy to 2025 (2020). https://www.ema.europa.eu/en/about-us/how-we-work/regulatory-science-strategy. Accessed 19 Mar 2025

European Medicines Agency (2025). https://www.ema.europa.eu/en/homepage. Accessed 19 Mar 2025

FDA (2024) Tissues and Tissue Products. https://www.fda.gov/vaccines-blood-biologics/tissue-tissue-products. Accessed 19 Mar 2025

FDA Focus Areas of Regulatory Science Report (2022). https://www.fda.gov/science-research/advancing-regulatory-science/focus-areas-regulatory-science-report. Accessed 19 Mar 2025

Fierce Pharma (2023). https://www.fiercepharma.com/special-reports/priciest-drugs-2023. Accessed 19 Mar 2025

Food and Drug Administration (FDA) (2025). https://www.fda.gov. Accessed 19 Mar 2025

Fraser of Allander Institute (2021) The contribution of medical research funding by charities to the UK economy. https://fraserofallander.org/wp-content/uploads/2022/03/FAI-The-contribution-of-medical-funding-by-charities-to-the-UK-economy-1.pdf. Accessed 19 Mar 2025

Frearson J, Wyatt P (2010) Drug discovery in academia- the third way? Expert Opin Drug Discov 5: 909–919. https://doi.org/10.1517/17460441.2010.506508. Accessed 19 Mar 2025

Global Market Insights (2023). https://www.gminsights.com/industry-analysis/contract-research-organization-cro-market#:~:text=Global%20market%20for%20contract%20research,rise%20in%20R%26D%20investments%20worldwide. Accessed 19 Mar 2025

Goate A et al (1991) Segregation of a missense mutation in the amyloid precursor protein gene with familial Alzheimer's disease. Nature 349:704–706. https://doi.org/10.1038/349704a0. Accessed 19 Mar 2025

Grassano N et al (2024) Exploring the global landscape of biotech innovation – Preliminary insights from patent analysis. Publications Office of the European Union. https://data.europa.eu/doi/10.2760/567451. Accessed 19 Mar 2025

Hampson G, Steuten L (2024) Netflix and pill: is there a role for volume-delinked subscription-style payments beyond antimicrobials? Expert Rev Pharmacoecon Outcomes Res 24:1–3. https://doi.org/10.1080/14737167.2023.2271171. Accessed 19 Mar 2025

Hodgson J (2024) Drug pipeline 4Q23 – gene editing delivers. Nature Biotech 142:167–169. https://doi.org/10.1038/s41587-024-02125-2. Accessed 19 Mar 2025

ICH E2E Pharmacovigilance Planning (Pvp) (2025). https://www.ema.europa.eu/en/documents/scientific-guideline/international-conference-harmonisation-technical-requirements-registration-pharmaceuticals-human-use-topic-e-2-e-pharmacovigilance-planning-step-5_en.pdf. Accessed 19 Mar 2025

ICH Guidelines (2025). https://ich.org/page/ich-guidelines. Accessed 19 Mar 2025

ICH M4: The Common Technical Document (2025). https://www.ich.org/page/ctd. Accessed 19 Mar 2025

ICMRA (2025). https://www.icmra.info/drupal/en. Accessed 19 Mar 2025

International Society for Pharmacoeconomics and Outcomes Research, Inc (ISPOR) (2025). https://www.ispor.org/. Accessed 19 Mar 2025

Investopedia (2024). https://www.investopedia.com/terms/v/venturecapital.asp. Accessed 19 Mar 2025

IPRP (2025). https://www.iprp.global/home. Accessed 19 Mar 2025

IQVIA (2022). https://www.iqvia.com/insights/the-iqvia-institute/reports-and-publications/reports/emerging-biopharma-contribution-to-innovation. Accessed 19 Mar 2025

IQVIA Global Use of Medicines Outlook to 2028 (2024). https://www.iqviainstitute.org. Accessed 19 Mar 2025

IQVIA Institute for Human Data Science (2024) Global Trends in R&D. Activity, Productivity, and Enablers. https://www.iqviainstitute.org. Accessed 19 Mar 2025

IQVIA Quarterly Pharmaceutical Market Outlook (2023). https://www.iqvia.com/-/media/iqvia/pdfs/files/iqvia-quarterly-pharmaceutical-market-outlook%2D%2D-november-2023.pdf. Accessed 19 Mar 2025

Kessel M, Vickrey C (2024) Why Japan lacks a vibrant biotech industry. Nat Biotechnol 42:694. https://doi.org/10.1038/s41587-024-02227-x. Accessed 19 Mar 2025

Knox R, Curfman G (2022) The Humira patent thicket, the Noerr–Pennington doctrine and antitrust's patent problem. Nat Biotechnol 40:1761–1763. https://doi.org/10.1038/s41587-022-01583-w. Accessed 19 Mar 2025

Kovalainen M (2020) A typical workflow in a preclinical pharmacokinetic experiment. Admescope. https://admescope.com/2020/02/25/a-typical-workflow-in-a-preclinical-pharmacokinetic-experiment/. Accessed 19 Mar 2025

Ku AF et al (2024) Reversible male contraception by targeted inhibition of serine/threonine kinase 33. Science 384:885–890. https://www.science.org/doi/full/10.1126/science.adl2688. Accessed 19 Mar 2025

Mallapaty S (2024) Cheaper versions of blockbuster obesity drugs are being created in India and China. Nature 630:797–798. https://doi.org/10.1038/d41586-024-02044-x. Accessed 19 Mar 2025

Martin L et al (2017) How much do clinical trials cost? Nat Rev Drug Discov 16:381–382. https://doi.org/10.1038/nrd.2017.70. Accessed 19 Mar 2025

Miller KL et al (2022) Drugs and biologics receiving FDA orphan drug designation: an analysis of the most frequently designated products and their repositioning strategies. Expert Opin Orphan Drugs 9:265–272. https://doi.org/10.1080/21678707.2021.2047021. Accessed 19 Mar 2025

Moon S, Erickson E (2019) Universal medicine access through lump-sum remuneration – Australia's approach to hepatitis C. N Engl J Med 380:607–610. https://doi.org/10.1056/NEJMp1813728. Accessed 19 Mar 2025

Mordor Intelligence: Contract Manufacturing Report (2024). https://www.mordorintelligence.com/industry-reports/pharmaceutical-contract-manufacturing-organization-cmo-market/market-size. Accessed 19 Mar 2025
Mordor Intelligence: Laboratory Equipment and Disposable Market (2025). https://www.mordorintelligence.com/industry-reports/laboratory-equipment-and-disposable-market. Accessed 19 Mar 2025
Mullard A (2024a) Experimental obesity drug packs double punch to reduce weight. Nature 629: 733–734. https://doi.org/10.1038/d41586-024-01433-6. Accessed 19 Mar 2025
Mullard A (2024b) 2023 FDA approvals. Nat Rev Drug Discov 23:88–95. https://doi.org/10.1038/d41573-024-00001-x. Accessed 19 Mar 2025
NIH History (2025). https://www.nih.gov/about-nih/who-we-are/history. Accessed 19 Mar 2025
NPMA Department of Drug Registration (2019). https://english.nmpa.gov.cn/2019-07/18/c_377593.htm. Accessed 19 Mar 2025
Orphanet (2025). https://www.orpha.net/. Accessed 19 Mar 2025
Pfizer: See How Biosimilars Are Developed (2021). https://www.pfizerbiosimilars.com/biosimilars-development. Accessed 19 Mar 2025
Pharmaceuticals and Medical Devices Agency (PDMA) (2025). https://www.pmda.go.jp/english/index.html. Accessed 19 Mar 2025
PharmExec.com (2020). https://www.pharmexec.com/view/myth-skyrocketing-drug-prices-closer-look-us-gross-net-problem. Accessed 19 Mar 2025
Pugsley MK et al (2008) Principles of safety pharmacology. Br J Pharmacol 154:1382–1399. https://doi.org/10.1038/bjp.2008.280. Accessed 19 Mar 2025
Ramakanth D et al (2021) Advanced packaging for distribution and storage of COVID-19 vaccines: a review. Environ Chem Lett 19:3597–3608. https://doi.org/10.1007/s10311-021-01256-1. Accessed 19 Mar 2025
Remmel A (2022) Why chemists can't quit palladium. Nature 606:448–451. https://doi.org/10.1038/d41586-022-01612-3. Accessed 19 Mar 2025
Rosenburg B et al (1965) Inhibition of cell division in Escherichia coli by electrolysis products from a platinum electrode. Nature 205:698–699. https://doi.org/10.1038/205698a0. Accessed 19 Mar 2025
Science the Endless Frontier 75th Anniversary Edition (2020). https://www.nsf.gov/about/history/EndlessFrontier_w.pdf. Accessed 19 Mar 2025
Shen J et al (2019) Design and conduct considerations for first-in- human trials. Clin Transl Sci 12: 6–19. https://doi.org/10.1111/cts.12582. Accessed 19 Mar 2025
Statista (2024). https://www.statista.com/statistics/791340/pharmaceutical-companies-number-with-active-pipeline/. Accessed 19 Mar 2025
The King's Fund (2023). https://www.kingsfund.org.uk/insight-and-analysis/data-and-charts/nhs-budget-nutshell#where-does-the-money-go. Accessed 19 Mar 2025
Verdin P (2024) Top companies and drugs by sales in 2023. Nat Rev Drug Discov 23:240. https://doi.org/10.1038/d41573-024-00041-3. Accessed 19 Mar 2025
Wadman M (2023) FDA no longer has to require animal testing for new drugs. Science 379:127–128. https://doi.org/10.1126/science.adg6264. Accessed 19 Mar 2025
Wagner J et al (2018) A dynamic map for learning, communicating, navigating and improving therapeutic development. Nat Rev Drug Discov 17:150. https://doi.org/10.1038/nrd.2017.217. s
Warne N et al (2023) Delivering 3 billion doses of Comirnaty in 2021. Nat Biotechnol 41:183–188. https://doi.org/10.1038/s41587-022-01643-1. Accessed 19 Mar 2025
WHO International Classification of Diseases (2025). https://icd.who.int/en. Accessed 19 Mar 2025
Wouters OJ et al (2020) Estimated research and development investment needed to bring a new medicine to market, 2009-2018. JAMA 323:844–853. https://doi.org/10.1001/jama.2020.1166. Accessed 19 Mar 2025
Zanders ED (2020) The science and business of drug discovery. Springer Nature, Switzerland AG. https://doi.org/10.1007/978-3-030-57814-5. Accessed 19 Mar 2025
Zhang Z, Tang W (2018) Drug metabolism in drug discovery and development. Acta Pharm Sin B 8:721–732. https://doi.org/10.1016/j.apsb.2018.04.003. Accessed 19 Mar 2025

Computational Approaches to Drug Discovery

3

3.1 Introduction

Our lives have been noticeably affected by computers in the last 30 years or so and this situation can only increase in unforeseeable ways. This point is particularly relevant to artificial intelligence and machine learning, a branch of computing that has only really entered public consciousness since the 2010s.

The history of computing in general has been well documented, for example in the timeline provided by the Centre for Computing History in Cambridge, UK (see Centre for Computing History 2025). Some (randomly selected) dates are as follows:

1605	Francis Bacon first uses binary (a code formed with a and b)
1613	Richard Braithwaite first uses the word *computer*
1703	Gottfried Wilhelm Leibnitz invents the binary system the 0 and 1 system
1873	Christopher Sholes invents QWERTY keyboard
1889	Nintendo is founded
1925	Julius Lilienfeld patents the transistor
1936	Alan Turing invents the principle of the modern computer
1947	First computer *bug* is found
1958	Jack Kilby creates the first integrated circuit
1964	Douglas Engelbart invents the mouse
1971	Ray Tomlinson sends the first email
1975	Bill Gates and Paul Allen found Microsoft
1976	Steve Wozniak, Steve Jobs, and Ronald Wayne found Apple Computer Inc.
1983	Microsoft announces Windows
1991	Tim Berners-Lee publishes the first ever website
1998	Larry Page and Sergey Brin found Google
2004	Mark Zuckerberg launches *Thefacebook* from Harvard dorm
2005	First video uploaded to YouTube
2012	Google Brain recognises cats in YouTube videos

(continued)

E. D. Zanders, *Twenty-first Century Drug Discovery*,
https://doi.org/10.1007/978-3-031-93949-5_3

1605	Francis Bacon first uses binary (a code formed with a and b)
1613	Richard Braithwaite first uses the word *computer*
1703	Gottfried Wilhelm Leibnitz invents the binary system the 0 and 1 system
1873	Christopher Sholes invents QWERTY keyboard
1889	Nintendo is founded
1925	Julius Lilienfeld patents the transistor
1936	Alan Turing invents the principle of the modern computer
1947	First computer *bug* is found
1958	Jack Kilby creates the first integrated circuit
1964	Douglas Engelbart invents the mouse
1971	Ray Tomlinson sends the first email
1975	Bill Gates and Paul Allen found Microsoft
1976	Steve Wozniak, Steve Jobs, and Ronald Wayne found Apple Computer Inc.
1983	Microsoft announces Windows
1991	Tim Berners-Lee publishes the first ever website
1998	Larry Page and Sergey Brin found Google
2004	Mark Zuckerberg launches *Thefacebook* from Harvard dorm
2005	First video uploaded to YouTube
2012	Google Brain recognises cats in YouTube videos
2019	Google claims to achieve quantum supremacy
2022	OpenAI launches ChatGPT

Of course, the impact of computing on drug discovery and development has been significant in many ways. This chapter will briefly outline the computational approaches used for the different stages of the drug discovery pipeline in order to provide a background to the specific examples given in the rest of the book. It will also describe some of the hardware and software tools being employed, as well as a basic description of the artificial intelligence and machine learning approaches being applied to biopharma R&D.

3.1.1 Drug Discovery Problems Addressable with Computers

Figure 3.1 shows an outline of the drug discovery and development process where computer analysis is, or will in the future, increase efficiency and possibly identify novel interventions without human input.

The different parameters (diseases, genes, lifestyle etc) relate to human subjects either selected at random from the general population, or with specific diseases. Extracting target and diagnostics information from this information is driving interest in machine learning where targets and interventions may become apparent without input of human knowledge (see later). Computation plays a part in all aspects of small molecule and advanced therapies, with preclinical development benefiting from accurate prediction of pharmacokinetic and safety issues. The conduct of clinical trials will be improved through the use of sophisticated patient stratification criteria and virtual control arms to minimise the use of placebos or active comparators.

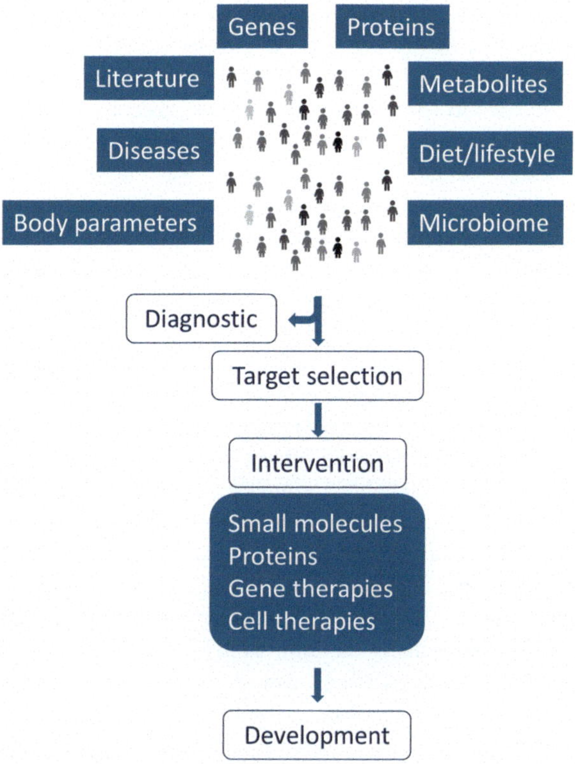

Fig. 3.1 Areas of drug discovery and development where computer analysis has, or is likely to have, a significant impact

3.2 Computing in the Biomedical Sciences

3.2.1 Molecular and Structural Biology

Molecular biology emerged as a discipline in the last half of the twentieth century and underpins much current research into drug targets and mechanisms of disease. The field is concerned with two concepts, information and structure (Stent 1968), the former relating to genetic information in the form of DNA, RNA and its specification of proteins (including drug targets). The latter concerns the three-dimensional structures of these macromolecules and is often described a structural biology. In each case the development of computers created the tools to replace laborious manual methods.

3.2.1.1 Structure Determination

Determining the three-dimensional (3-D) structures of small molecules and macromolecules such as DNA and proteins was made possible by the development of X-ray crystallography. The famous 3-D structure of DNA was published in 1953 based on an intuitive approach by Watson and Crick to create a model based on all available experimental data including the simple helical structure inferred from X-ray data. X-ray patterns have to be converted into electron density maps using the mathematical process of Patterson and Fourier transformations, and computers were used to speed up these laborious tasks by John Kendrew, a pioneer in protein structure determination. As described by de Chadarevian (2018), Kendrew's research student Hugh Huxley was tasked in 1949 with calculating a two-dimensional Patterson projection of one of the haemoglobin forms, a tedious operation requiring about two weeks of solid work. John Bennett, working on EDSAC, the pioneering electronic computer with a stored memory program that had just started operating, managed to produce a programme which could do Huxley's two weeks' work in about half an hour. Another half an hour was needed to print out the results. Kendrew learnt programming himself and together they perfected the programme that was published as a joint paper in 1952. It was the first paper written on crystallographic computations using electronic digital computers.

Early computers like EDAC were unreliable and needed specialist skills to operate, but inevitably, progress was made in the decades following Kendrew's work. Attempts to predict the conformation of proteins from primary amino acid sequence alone *the protein folding problem* could be undertaken with new hardware, programming languages and a theoretical understanding of the potential energy of a molecule. As described in a reminiscence by Levitt (2001) the total potential energy of any molecule is the sum of terms allowing for bond stretching, bond angle bending, bond twisting, van der Waals interactions and electrostatics. Many properties of a biomolecules, including proteins and nucleic acids, can be simulated with such an empirical energy function. In 1967, Levitt used a computer at the Weizmann Institute in Israel (called the Golem, after the Jewish folklore automaton) to develop a programme CFF, to calculate structural, thermodynamic and spectroscopic properties of proteins; this led to the first energy minimization of an entire protein structure in a process that became known as *energy refinement*.

Computational structural biology developed further and used, for example, in homology modelling where proteins without an experimentally determined 3-D structure could be compared with the known structures populating databases such as the protein data bank (PDB). With advances in machine learning, it is now possible to predict the 3-D structures of proteins without a reference structure, a major step forward to be discussed later.

3.2.1.2 Sequence Analysis

Without the means of determining the sequence of nucleotides in DNA and RNA, or amino acids in proteins, there would be no modern biotechnology. Much of the credit in this area can go to Fred Sanger who won two Nobel prizes in chemistry, the first for his sequencing of insulin in 1955 and the second for DNA sequencing in 1975 (see Brownlee 2013). Sequencing technology has advanced considerably since then as will be covered later, but the sequencing data emerging as a result of the work of Sanger and others prompted the development of bioinformatics. Sequence analysis using computers started with proteins, since Sanger's work with insulin and Edman's modifications preceded DNA sequencing by some years. Margaret

Dayhoff was a pioneer in this field such that she was called *the mother and father of bioinformatics* (Gauthier et al. 2019). She and Robert Ledley developed the first bioinformatics programme COMPROTEIN from 1958 to 1962 to process Edman sequencing data, introducing in the process the three letter amino acid codes for proteins. Dayhoff later devised the one letter amino acid code making it easy for sequences to be represented as text strings. The same could be said of DNA analysis, with the four bases represented as A,C,G, and T, the last substituted with U in RNA. With the genetic code deciphered, it was possible to work out the amino acid sequence of proteins encoded in DNA by using computer software as well as being able to to compare large numbers of DNA sequences in parallel. However, none of this could reach its full potential until the computing hardware was reduced in size and increased in power.

The first software collection for sequence analysis was GCG developed by the University of Wisconsin Genetics Computer Group in 1984. This suite of programmes ran on a minicomputer, but the introduction of microcomputers in the 1980s stimulated the further development of bioinformatics software that could be run on smaller machines. Of note was the move to free, open-source software allowing the expansion of DNA/protein analysis tools and the establishment of sequence databases from the 1980s onwards. The beginnings of whole genome sequencing in the 1990s provided the impetus for more software development aided by the invention of the World Wide Web by Tim Berners-Lee. Sequence databases like the EMBL Nucleotide Sequence Data Library hosted online in 1993 and in 1994, the NCBI database which introduced the sequence comparison tool BLAST.

As the amount of nucleotide and protein sequence data accumulates worldwide there is a pressing need for consolidated and easily searchable databases. By analogy to the Google search engine, a software tool called Metagraph has been developed to compress over 50 petabases of sequence data from the NCBI's sequence read archive to make it easily searchable without expensive hardware (see Offord 2024 for background and discussion of similar initiatives).

3.2.2 Chemoinformatics and Virtual Screening

Proteins may represent the majority of drug targets, but it is their interaction with other molecules that is fundamental to drug discovery. This section provides some background to the computational tools developed to measure these interactions for identifying suitable ligands as drug candidates or designing them *de novo*.

3.2.2.1 Chemoinformatics

The term *chemoinformatics* was introduced in 1998 by Dr. Frank K. Brown as a generic term encompassing the design, creation, organization, management, retrieval, analysis, dissemination, visualization and use of chemical information (Chemoinformatics in Drug Discovery 2005). Related terms include cheminformatics, chemi-informatics, chemometrics, computational chemistry, chemical informatics, and chemical information management/science.

As with DNA and protein sequences, chemical structures have to be converted into a machine-readable format to be interpreted by computers. The *structure data* (SD) file is a store of chemical structures using bond lengths to specify x and y atom coordinates, z coordinates for 3-D representation, and atom types, chirality and bond types. The SD file is a connection table based on chemical graph theory in which the edges represent the covalent bonding between atoms and the vertices represent atoms (hydrogen atoms are omitted). It is essentially a graph containing a full description of the molecular topology to be used in virtual screening and other applications. Storage becomes an issue when using large databases of virtual compounds at the terabyte (TB) level, so file types such as the *Simplified Molecular Input Line Entry System* (SMILES) have been introduced that use 50–70% less space than other formats (Good 2007). The rules for generating these strings can be found in EPA SMILES Tutorial (2025). Optimal encoding of chemical structures in computer-readable forms is important enough for standard machine reading algorithms but becomes critically important for deep learning applications outlined later.

Examples of graph rendering of cimetidine (Li et al. 2018) and the SMILES format for morphine (PubChem 2025) are shown in shown in Fig. 3.2.

3.2.2.2 Virtual Screening

The screening and selection of compounds *in silico* has an obvious attraction in terms of efficiency and throughput. Once suitable candidates have been identified they can be then be synthesised and tested in the lab. The sheer number of virtual compounds that could be made increases the chance of finding novel chemotypes, possibly leading to improved drug molecules. The success of this approach depends upon the computational techniques used, particularly for *structure-based virtual screening* (SBVS). This procedure is designed to predict the interaction mode between two molecules joined in a stable complex by using *scoring functions* to estimate the force of non-covalent interactions between a ligand and molecular target. Search algorithms are used in conjunction with scoring functions to search for ligand orientations and conformations at the binding site. To predict the correct conformation of ligands, the algorithms check the chemistry and geometry of the atoms involved and the rotational or conformational degrees of freedom in the molecule's movement (Maia et al. 2020).

Scoring functions determine the ligand binding site and the conformation between a target and ligand and predict the binding affinity between a protein and ligand. As well as being used for screening of virtual libraries, the scoring function can also be used for the important process of lead optimisation. Figure 3.3 illustrates the use of a scoring function to select an active compound (celecoxib) binding to target enzyme (cyclooxygenase 2) compared with an inactive entity. In (a), the calculated binding affinity of the inactive compound is approximately half that of celecoxib in (b). Also shown are the various interactions between the small molecule and amino acid residues in the target protein used to derive the affinities.

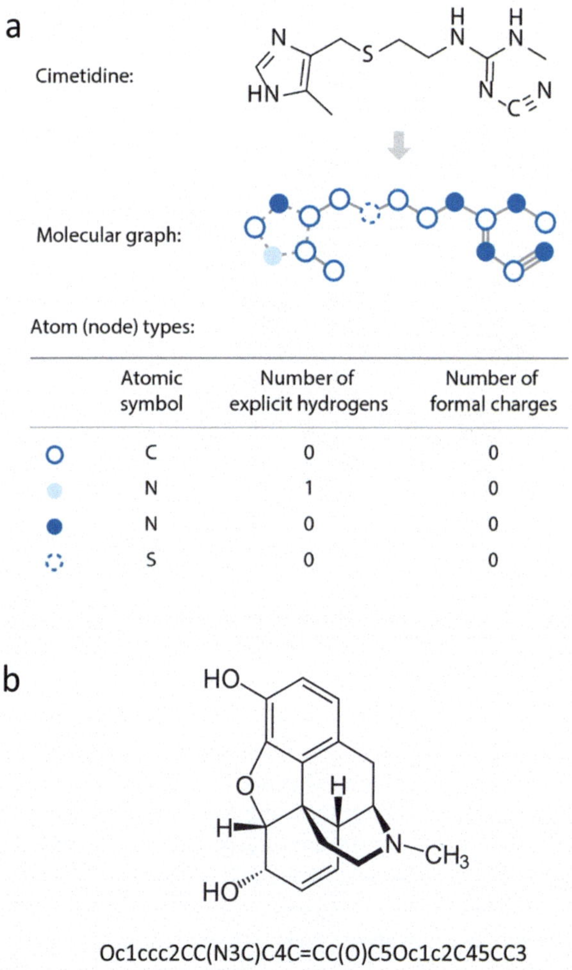

Fig. 3.2 (**a**) Molecular graph rendering of cimetidine. (From Li et al. 2018 under CC- BY-4.0). (**b**) SMILES string rendering of morphine. (PubChem 2025)

It is important to evaluate the quality of the docking process, the most common methods being *root mean square deviation* (RMSD) and *receiver operating characteristic area under the curve* (AUC-ROC). RMSD calculations evaluate whether a docking programme can reliably reproduce a known crystallographic conformation as well as their respective intramolecular interactions. Ligand-target complexes are redocked and the average distance between atoms in the known ligand and the virtual one (obtained *in silico*) are calculated as the RMSD distance. Values $\leq$2.0 Å are considered acceptable.

A universal problem in small molecule screening is the discrimination between *false positive compounds* (FPCs) and *true positive compounds* (TPCs), something that is particularly acute when screening enormous virtual libraries. The ROC and AUC-ROC methods use distributions of binding energy observed when using TPCs and a set of decoys. The latter are compounds that are inactive against the target but possess similar properties to the former, for example in molecular mass and the

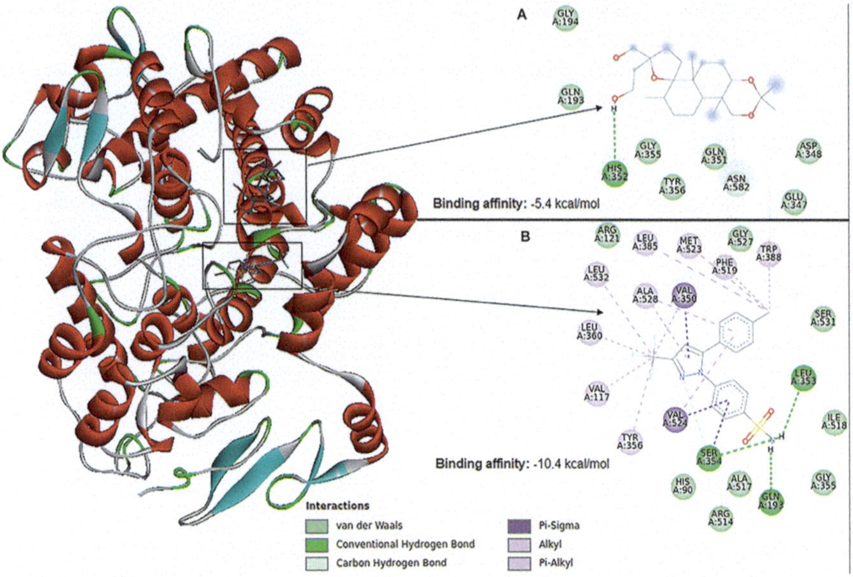

Fig. 3.3 Calculating the affinity of binding between cyclooxygenase2 and an inactive ligand (**a**) or the active molecule celecoxib (**b**). (Figure taken from Maia et al. 2020 under CC-BY-4.0)

number of rotatable bonds. TPCs are expected to have lower binding energies, and the ROC curve plots the distribution of true and false results on a graph (see Narkhede (2018) for an illustration of the basic principles). The AUC-ROC reflects the probability of discriminating between active and inactive compounds. The closer the AUC is to 1, the greater discrimination between TPCs and FPCs, values close to 0.5 indicate a random process and acceptable values should be >0.7.

Needless to say, the whole area of virtual screening is more involved than the summaries given above, and it is evolving, particularly with the influence of machine learning. Some background to these approaches and their potential for future drug discovery can be found in Sadybekov and Katritch (2023).

3.2.2.3 Data Sources

A small selection of the main data sources for bio-and chemoinformatics is shown in Table 3.1.

The protein structure database entry includes the AlphaFold Protein Structure database produced using machine learning and containing over 200 million predicted structures. This will be covered in more depth later in the book.

3.2.3 Systems Biology

The bulk of drug discovery research involves a *bottom-up* approach where individual proteins, or other molecules, are isolated from the cellular environment as

Table 3.1 List of bioinformatics and chemoinformatics databases

	Database	URL
DNA	INSDC	https://www.insdc.org/about-insdc/
	GenBank	https://www.ncbi.nlm.nih.gov/genbank/about/
	European Nucleotide Archive	https://www.ebi.ac.uk/ena/browser/home
	DDBJ	https://www.ddbj.nig.ac.jp/index-e.html
Protein	UniProt	https://www.uniprot.org/
	InterPro	https://www.ebi.ac.uk/interpro/
Structure	RCSB PDB	https://www.rcsb.org/
	AlphaFold protein structure database	https://alphafold.ebi.ac.uk/
Compounds	PubChem	https://pubchem.ncbi.nlm.nih.gov/
	ChEBI	https://www.ebi.ac.uk/chebi/
	CAS registry	https://www.cas.org/cas-data/cas-registry
Virtual libraries	Enamine	https://enamine.net/compound-collections/real-compounds/real-database
	ZINC database	https://wiki.docking.org/index.php?title=Main_Page

potential targets for new drugs. However, targets rarely act in isolation, they are part of a complex interconnecting system that may be better understood through a *top-down* approach of looking at the intact cell/tissue and working downwards towards an actionable molecule or process. This can be addressed by using computational tools under the umbrella term *systems biology*, introduced in 1968. The field took some time to be accepted by the scientific community who had difficulty with the idea that complex biology could be modelled using mathematics (see Voit 2022 for background). Although his problem still exists, there are clearly areas of drug discovery that will benefit from a more holistic approach to target discovery and modelling of physiology at the cell and tissue level for applications in preclinical development. This has been discussed by Butcher et al. (2004) some detail.

Currently the rapid implementation of artificial intelligence and machine learning in R&D is providing a major impetus for the future development of systems biology and one day it may be possible develop all drug candidates without recourse to animal experiments (see Chap. 7).

3.2.4 Artificial Intelligence and Machine Learning

3.2.4.1 Computer, Make Me a Drug

Imagine a future in which a human drug discovery scientist (with plenty of leisure time) asks a computer and robotic surrogates to produce a medicine for a specific disease and patient population. Massive datasets are consolidated and processed, drug candidates are selected based on a high probability of success in preclinical development and clinical trials, and all conducted without human intervention except finally in patients. This may be fanciful perhaps, but current biopharma R&D is moving in this direction, even if the whole process (illustrated in Fig. 3.1) is in its infancy and not fully joined up.[1] It is possible to break these processes down into individual elements where AI is, or can have, an impact. AI in the context of this introduction is a catch-all term for *artificial intelligence* and *machine learning*, to be described later in some detail.

Diseases and Drug Targets

Diseases are often heterogeneous, cancer being a good example, and initiated or progressed through genetics, the environment and the microbiome etc.[2] Knowledge about particular diseases has been gained over many years by physicians and scientists and published in the biomedical literature or presented at conferences etc. As a result, the relevant material is often the first port of call when thinking about a drug discovery project in a particular disease area. Until the advent of the internet, scientific publications used print media, first introduced in the seventeenth century (The Philosophical Transactions of the Royal Society was first published in 1665). According to Ghasemi et al. (2023) there are 52,564 active journals covering biomedical sciences of which 13,928 journals are indexed by the Institute for Scientific Information (ISI) and 1960 journals have an impact factor (IF) of zero. They quote an estimated 3.4 million articles published in 2020. The PubMed database from the National Center for Biotechnology Information (NCBI) has been publicly available online since 1996 and contains over 37 million citations. Clearly this is a massive amount information which given its scale, will be of variable quality. Nevertheless this provides ample opportunity for electronic mining of an expanding biomedical literature at a scale that no human operator could possibly achieve. There are also many other sources of information that could be used collectively to analyse diseases and identify potential drug targets. This information includes DNA, RNA and protein sequence and expression data (including the microbiome and epigenome), disease associated mutations and other parameters, metabolite profiles, body parameters, diet and lifestyle. The key to using this information for AI/machine learning lies in the curation of large databases containing good quality data that can be converted into a machine-readable format. The databases relating to genetic information will be covered in Chap. 5, but to give one example, the UK Biobank has since 2006, collected data from 500,000 people as part of a large-scale prospective study. Analysis of their blood, urine and saliva samples as well as detailed information about their lifestyle and environmental parameters provides a comprehensive resource that is freely available for researchers worldwide (UK Biobank 2025). AI has the potential to identify novel drug targets and diagnostics through mining these different sources of data, including

[1] 'AI Scientist' is an effort to do the joining up, but is at a very preliminary stage (see Castelvecchi 2024)

[2] Syndromes are collections of associated signs and symptoms but without a clear cause (aetiology)

publications. Many reviews on this subject are appearing in the literature, with Pun et al. (2023) being a relevant example.

Drug Candidates
Small molecule drug discovery involves chemical synthesis and structural evaluation along the lines described in Sect. 3.2.2 above. AI finds a ready application in these areas as well as large molecule discovery. These applications will be highlighted in Chap. 6.

Preclinical and Clinical Development
AI is well suited for predicting various outcomes in preclinical development such as pharmacokinetics, toxicity and formulation; this is because machine learning methods can be used to identify sets of criteria that could be applied to novel compounds. Clinical development requires both the selection of relevant disease parameters and the subjects likely to respond to a given treatment. These are both areas where AI can be applied, along with rapid and accurate monitoring of clinical outcomes using, among other things, image analysis. As with the previous categories in this section, further elaboration of the use of AI in drug discovery and development will be highlighted where appropriate; of course each category has applications in common which will become apparent.

3.2.4.2 Background to AI

Most people are by now aware of the existence of artificial intelligence (AI) and its machine learning offshoots. This field has ramifications far outside the world of drug discovery and development and is beginning to impact upon daily life in different ways. Chatbots, online marketing and even the voicing of actors in TV crime reconstructions are all originating from AI algorithms; however discussions about the positive and negative aspects of AI and its implementation in multiple fields is outside the scope of this book. Since the AI technology relevant to drug discovery and development is expanding rapidly it is likely to underpin a significant amount of R&D in the biopharma industry over the coming decades. To make a general point, the number of software engineers worldwide is projected to rise to 45 million by 2030 (Uspenskyi 2024) and it is speculated that around 60,000 engineers work for Google alone.

AI is a mathematical process based on statistics that requires advanced computer hardware and software, often on a massive scale. This section provides an overview of the relevant AI technology in an attempt to convey the basic details in a form that should be comprehensible to a non-specialist. Of course, it is only possible to really understand what is happening in AI *under the hood* by studying the relevant algorithms and computer programming using tools like Python, but this introduction will hopefully give a flavour of what is involved.[3] Luckily there is no shortage of

[3] Even then, AI is often seen as a *black box* where it is not clear how the end results are actually generated

online and other resources for those wanting more detailed information. A comprehensive analysis of AI and its impact can be found in the Stanford University AI Index Report (2024). For more specific coverage of AI in drug discovery and development see the Wellcome Trust Report (2023).

Origins of Artificial Intelligence

The origin of the term *Artificial Intelligence* can be seen in Fig. 3.4 showing the plaque at Dartmouth College USA which commemorates a 1956 meeting in which the term was coined (Moor 2006).

The idea of mimicking the human brain by using a machine predated 1956, with pioneers such as Alan Turing, John Von Neumann and others establishing the foundations of computer science; however, the Dartmouth event did establish a new field that would develop further with solid theoretical foundations and improved computers.

The objective of AI is to replicate the functions of the human brain in a computer with the, as yet unachieved, goal of creating *Artificial General Intelligence* (AGI). It follows that at least a basic understanding how the brain works is an essential starting point. As an aside, the question could arise: whose brain? Would it be an Einstein, Shakespeare, or any number of people with remarkable powers of memory or

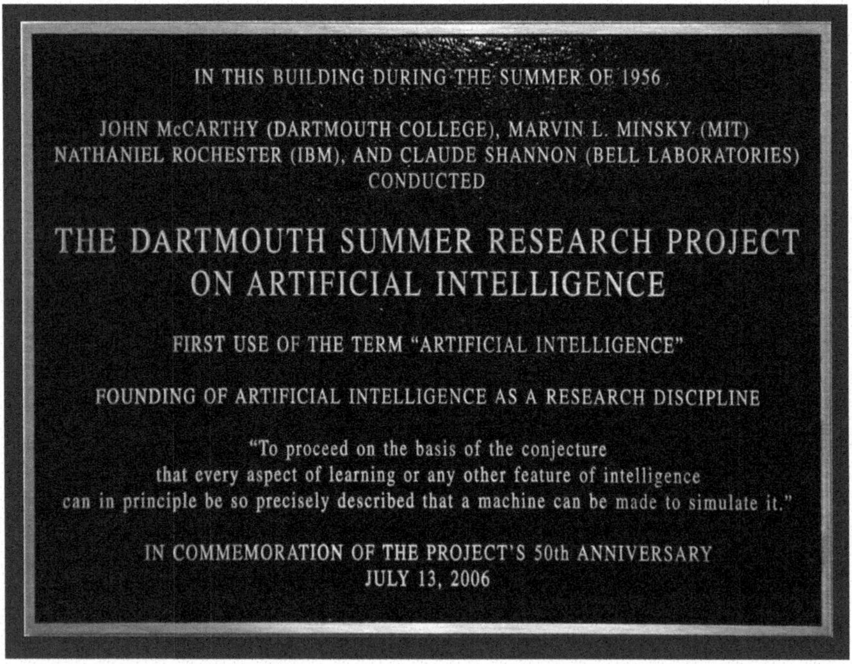

Fig. 3.4 Plaque commemorating origin of term *artificial intelligence*. (From Moor 2006 under CC-BY-4.0)

creativity? Regardless of the above, little is known about any brain, although the basic architecture of neurons and their connections are being mapped and studied in molecular detail. These connections (*connectomes*) in the human brain are currently too large to identify as a whole, but the small model organism *Caenorhabditis elegans* has only 302 neurones which are hardwired and readily amenable for study. The larger model organism *Drosophila melanogaster* has 140,000 neurons that are not hardwired in the same way as *C.elegans*, while mice have around 70 million compared with an estimated 86 billion in humans. It may however be possible one day to look at the issue in reverse and use AI to help in understanding how the biological brain works (Jain 2023).

3.2.4.3 Basic Concepts

Artificial Intelligence (AI) is a broadly based study of intelligent machines, or at least those with the appearance of having the same characteristics as human intelligence. Machine Learning (ML) is the part of AI that uses computers to learn from data, with a subgroup called Deep Learning (DL) based on *deep neural networks*. DL is currently the most active field of machine learning with many applications including in biopharma R&D. The above inter-relationships are summarised in graphical form in Fig. 3.5.

Machine learning algorithms are trained on data, learn from them, and develop a generalised understanding that allows computers to make predictions and decisions on new unseen data. The main uses for machine learning algorithms include:

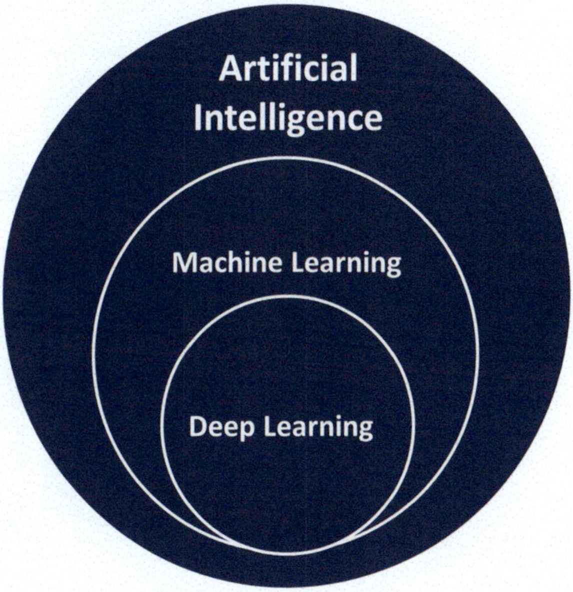

Fig. 3.5 Artificial Intelligence, Machine Learning and Deep Learning are separate but related entities

Supervised Learning algorithms input training data and generate training outcomes for comparison with optimal outputs specified by the user. The algorithm learns how to optimise the processes given these specifications.

Unsupervised Learning generally uses only inputs of unstructured training data to find patterns that can be analysed using statistical tools.

Semi-Supervised Learning uses unstructured data in combination with smaller portion supervised training data to accelerate the machine learning process.

A summary of commonly used algorithms useful for drug development (adapted from Dara et al. 2022) is shown in Fig. 3.6.

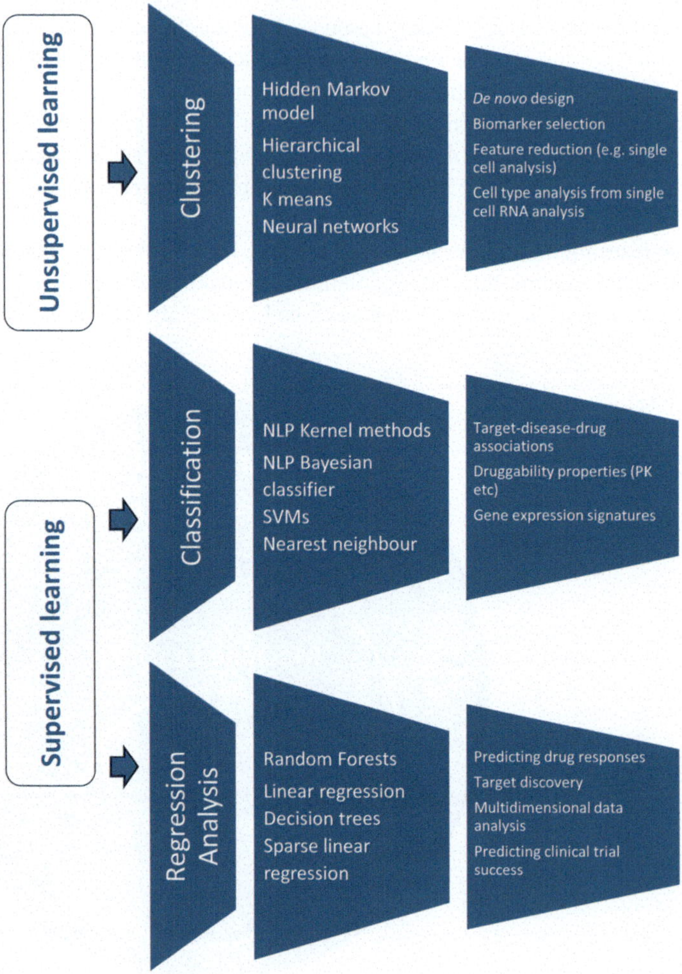

Fig. 3.6 Unsupervised and supervised machine learning algorithms. (Adapted from Dara et al. 2022)

Clustering algorithms generate associations between different data parameters using statistical methods. Those illustrated in Fig. 3.6 are:

Hidden Markov models
Hierarchical clustering
k-means
Neural networks
Classification algorithms, including NLP or *natural language processing*, to be discussed later
SVMs or Support Vector Machines
Nearest neighbour analysis
Regression Analysis algorithms include:
Random Forests
Linear regression
Decision Trees
Sparse Linear Regression

Elaboration of most of the above is beyond the scope of this book but of course there is plenty of information about each algorithm online and elsewhere. However, brief descriptions do follow for *decision trees*, *random forests* and *neural networks*, the last being central to deep learning algorithms.

Decision Trees

Decision Trees have been used for many years to split input data into decision nodes (where the data are split) and leaves (decisions/outcomes). A very simple example is given in Fig. 3.7 where different criteria are used to determine the risk of having a heart attack. Much more sophisticated trees can be used for selection of biomarkers and diagnostics and other investigations in drug discovery, but the basic principle illustrated here is the same for all.

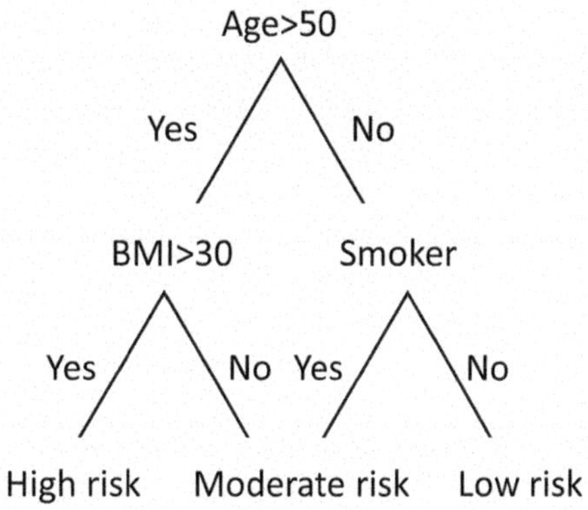

Fig. 3.7 A simplified binary decision tree to determine the risk of a heart attack. *BMI* body mass index

Decision trees are used with Supervised Machine Learning as Classification trees (predicted outcome is the same discrete class as the input data and give a yes/no answer) and Regression trees (where the predicted outcome is a real number). Classification and Regression tree analysis (CART) is used as an umbrella term for both procedures. The situation becomes more complicated when dealing with large datasets and the need to assess the reliability of the output data. As with the whole field of machine learning this is a complex area for those without backgrounds in statistics and computer science, but the basic principles are quite easy to understand if not the detailed practical implementation.

It is often not necessary (or feasible) to analyse every member of a sample used for supervised training, so the process of *bootstrapping* (or *bootstrap bagging*) is employed to analyse a manageable subset. Essentially, samples representing a small proportion of the dataset (bootstrap samples) are taken at random and used with the decision tree algorithm. *Out of the Bag* data (OOB) is the portion of the original population that has never been selected in any of the random samples. so testing it gives an estimate of the quality of the model. A value Q^2 is derived which gives a measure of the *predictive relevance* of the model, with $Q^2 > 0$ being good. Generating reliable data from decision tree algorithms is clearly critical when studying complex phenomena as encountered in drug discovery and elsewhere. *Ensemble methods* combine the predictions of several training estimators to provide a more robust result. One method is the *Random Forest* algorithm a collection of decision trees (ensembles) where each tree gets a *vote* in classifying the sample and find patterns in the data. Achargee et al. (2016) provide an example of using random forests to integrate metabolomics, lipidomics and clinical data relating to changes in the livers of rats treated with a PPAR agonist.

3.2.4.4 Deep Learning

A little Learning is a dangerous thing; Drink deep, or taste not the Pierian Spring.

This rhyming couplet from Alexander Pope's *Essay on Criticism*, although written in 1711, perhaps has a certain relevance to computer science 300 years later (*Pierian* refers to the Greek mountain home of the Muses).

Deep learning is based on the properties of *deep neural networks* that mimic the learning abilities of the human brain. Neural networks have been studied for many years and consist of a network of interconnected nodes containing input and output commands. They are capable of performing complex tasks through emergent functionality along the lines of *the whole is greater than the sum of its parts*. Much of the current interest in machine learning comes from engineering deep neural networks that use hidden decision-making layers in order to automatically learn hierarchical representations of data. As such, deep learning networks are used for image and speech recognition along with natural language processing (see later).

3.2.4.5 Deep Learning in Practice

Deep neural networks provide the foundation for algorithms that can perform a variety of functions such as image and text recognition using deep learning architectures mainly based on *Convolutional Neural Networks (CNNs)* and

Transformer models respectively. Early models were devised for specific tasks and needed substantial computational resources. More recently, *Foundation Models* based on *generative artificial intelligence* have been used to process a large corpus of untrained data for different applications while minimising computational resources. Generative AI creates data that did not previously exist using self-supervised and transfer learning to apply information it's learnt about one situation to another (IBM Research 2022). A list of foundation models is shown in Table 3.2.

Convolutional Neural Networks (CNNs)

CNNs are used for image recognition and provide a useful model for showing, up to a point, how deep neural networks operate. Figure 3.8 shows the basic architecture of a deep neural network. Each node is a *neuron* connected by *edges* by analogy with the biological equivalent of neurons and synapses.

One way of explaining how the networks work is to consider handwritten character recognition. The networked human brain can recognise a handwritten number like 6 despite it being unevenly presented compared with a printed version. The brain has

Table 3.2 Foundation models. The natural language processing descriptions summarise the ability of these models to interact with humans, as chatbots for example. Each model has its own performance and amount of training data input

Model	Description	Originator
BERT	Natural language processing	Google
Med-PaLM2	Medical interface	Google
Chat GPT4	Natural language processing	Open AI
DALL-E3	Image to text processing	Open AI
Stable diffusion 3	Text to image processing	Stability.AI
Claude 2	Natural language processing	Anthropic
Llama 3.2	Natural language processing	Meta
Mistral AI	Natural language processing	Mistral AI

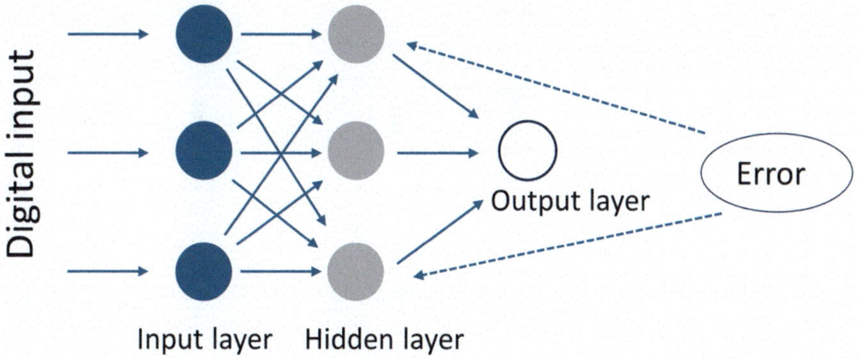

Fig. 3.8 Basic layout of deep neural network

learnt how the number should look and acts accordingly. Using 6 as an example, deep neural networks are designed to operate by creating a digital input of image pixels that are assigned values of 1 for black and 0 for white, with intermediate values in various levels of grey. In a process of *forward propagation*, the input layer passes the information to the first of possibly many hidden layers, (layers that are neither input or output). With each layer, the CNN increases in its complexity identifying greater portions of the image such as a loop and tail making up the character 6. As the image data progress through the layers of the CNN, it starts to recognize larger elements or shapes of the object until it finally identifies the intended object, in this case switching on the output layer identifying the 6. Inevitably an untrained network will produce errors that need to be identified and fed back to the model to apply corrections, something that will be described later. The strength of the connections between neurons is defined mathematically as a *weight*, while the strength required to activate a single neuron under specific input conditions is called a *bias*. Both of these features allow the network to learn by adjusting these parameters in a process of optimisation or *backpropagation* that measures the discrepancy between the expected output data and the actual input data, i.e. a *loss function* which is calculated in different ways and must be minimised. There is more than one way of calculating loss functions, each one being suited to a particular scenario as highlighted in Brooks (2024). As one example, *stochastic gradient descent* operates by calculating the gradient of the difference between predicted and actual values based on weights and biases, the idea being to identify the steepest error decrease to find a local minimum and achieve this by adjusting the weights and biases.

Figure 3.9 shows a single neuron with weighted inputs leading to an *activation function* that transforms the weight and bias values to a figure between 0 and 1 determining the connection between layers and ultimately the output.

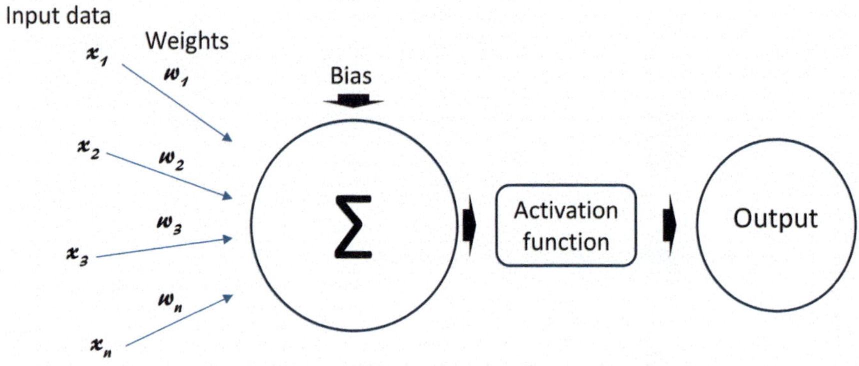

Fig. 3.9 Input to a single neuron modified by weights and biases. The summed weights and biases are converted through an activation function to a number ranging from 0 to 1 that will determine the strength of the output

The above field of deep neural networks is changing rapidly, making it even more complicated for those without a mathematics or computing background. Nevertheless there are some useful resources listed here that cover some of the basics:

H2O.ai Wiki (2025)
IBM Think (2025)
Weka.io (2025)
3Blue1Brown (2016)

Transformers

Transformers change an input sequence into an output sequence that is related in context. This could be English words translated into French. Unlike earlier neural networks, transformers use a *self-attention mechanism* to examine different parts of the input data sequence all at once and to determine the most important parts.[4]

The input sequence could be words, images, DNA sequences, sounds etc., but each must be converted into machine-readable forms known as *embeddings*. The sequences are broken down into different components (*tokens*) e.g. words in a sentence. Embeddings transform the token into a mathematical vector sequence representing multidimensional coordinates, in effect reducing the *dimensionality* of the data and helping ML models to find similarities among sparsely distributed items.

For example, a city could be represented by two coordinates: location (latitude/longitude) and population. By increasing this embedding to thousands of attributes about any word's grammar, meaning, and use in sentences, the transformer can calculate the relationships between words in mathematical terms and understand the human language model. Since the transformer must consider the order of the tokens in the input sequence a system of *positional encoding* adds information to each token's embedding to indicate its position in the sequence and understand the sequence context. See AWS: What are transformers in artificial intelligence? (2025) for a basic summary of transformer architecture.

Large Language Models (LLMs)

LLMs are generative AI algorithms trained on vast amounts of text data (*parameters*) to enable *natural language processing* (NLP). Models trained on more data will usually have more parameters and outperform those trained on less data.

NLP has many applications, including translation, chatbots and advanced human-computer interactions. The introduction of OpenAI's GPT (Generative Pre-trained Transformer)[5] series of LLMs has brought the potential of AI into the public domain

---

[4]Transformer concept entitled *Attention is all you need* (after Beatles' *Love is all you need*) published by Google scientists in 2017

[5]The DNN architecture for ChatGPT3 consists of 175B input parameters and 96 layers with 12,288 units in each of the latter (Brown et al. 2020)

(Open AI 2024) with a Chinese open-source version DeepSeek V3 introduced in 2025. The issues surrounding data acquisition for these models and the accuracy of output will be covered later.

The success of models like ChatGPT depend upon advances in transformers like Google's BERT (Bidirectional Encoder Representations from Transformers) which uses *masked language models (MLMs)* for pre-training. MLMs take a sentence, randomly masks some of the words, and then ask the model to predict these words based on the context provided by the unmasked words. Some input tokens (e.g. words) are masked at random and after repeated masking and prediction using large amounts of text, the most likely missing words can be assessed based on the highest probability.

Diffusion Models

Text to image systems can be used in a trivial way like *show me a cat jumping over the moon* where a generative AI model will show just that. Of course it has more serious applications, and like GPTs, can present the results of a complex analysis to with words and pictures a human operator in an intuitive way. Open AI's DALL.E models (1–3) are one example based on a *diffusion process*. Diffusion Models destroy training data through the successive addition of random Gaussian noise and then learn to recover the data by *reversing* this noising process. After training, the Diffusion Model generates data by simply passing randomly sampled noise through the learned denoising process. Algorithms based on this process have been developed for molecular docking e.g. Diffdock (Corso et al. 2023) but some questions about the accuracy of these algorithms for docking studies have been raised by Buttenschoen et al. (2024).

Knowledge Graphs

A biomedical *knowledge graph* (KG) is a resource that integrates one or more expert-derived sources of information into a graph where nodes represent biomedical entities and edges represent relationships between two entities (Nicolson and Greene 2020). KGs (also known as *semantic networks*) can be populated by manual input from various databases, but this is slow and inefficient process compared with machine learning approaches which can process large amounts of data automatically. Furthermore, natural language processing allows the investigator to query the graphical display and obtain an answer in text form, to quote Petrina Kamya from InSilico Medicine in Montreal: *If a scientist wants to investigate psoriasis, for example, the chat function can look at the knowledge graph for that disease. It will deliver a text description that includes the major signaling pathways and genes involved in psoriasis and the compounds known to interact with them* (Savage 2023).

An example of a knowledge graph based on biomedical data is shown in Fig. 3.10 (from Dörpinghaus et al. 2022).

The graph in Fig. 3.10 was built from PubMed metadata plus ontologies or terminologies like MeSH and is described in detail by Dörpinghaus et al. (2022). Processes can be described using Biological Expression Language (BEL) that

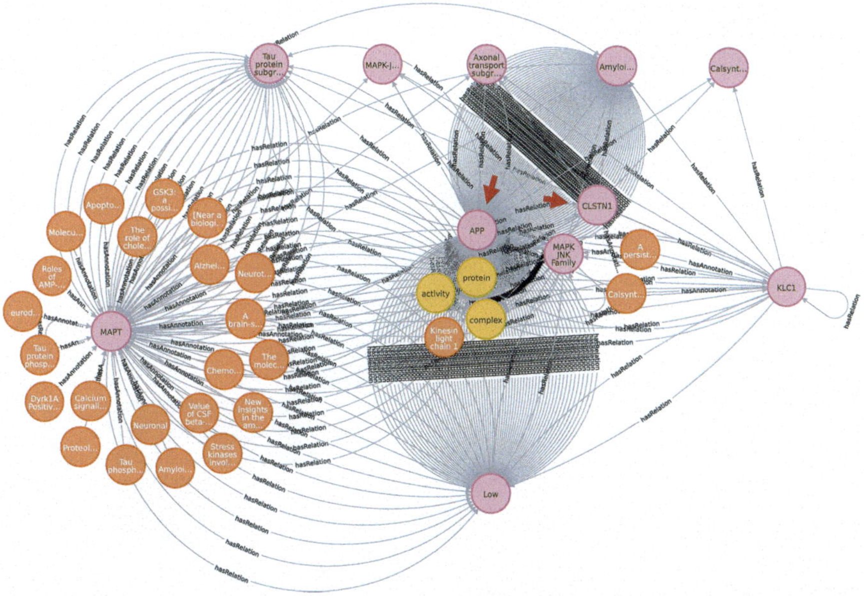

Fig. 3.10 Part of Knowledge Graph describing interactions leading to Alzheimer's disease. Orange circles: source documents; pink circles: context entities. Amyloid-beta precursor protein (APP), and Calsyntenin-1 (CLSTN1) are highlighted with an arrow. (Reproduced from Dörpinghaus et al. 2022 under CC-BY-4.0)

captures causal, correlative, and associative relationships between biological entities along with the experimental/biological context in which they were observed as well as the provenance of the publication from which the relation was reported (Biological Expression Language 2024). The KG in Fig. 3.10 uses the BEL statement act(p(HGNC:KLC1))–>p(HGNC:MAPT) to describe a key event in the disease process, namely the phosphorylation activity of human (HGNC) "Kinesin light chain 1" (KLC1) leading to phosphorylation of "Microtubule-associated protein tau" (MAPT).

The importance of mining the relationships between disparate sets of data in biomedical research has not been lost on the biopharmaceutical industry, with large companies like GSK and AstraZeneca fully engaged, along with informatics companies such as BioRelate (2025) who provide knowledge graph and other services to generate hypotheses for drug discovery and development.

Explainable AI (XAI)

Deep neural networks are seen as *black boxes* where it is not at all obvious how a particular algorithm, ChatGPT for example, actually arrives at a result. Being unable to *show the thinking* for solving a problem is not satisfactory for obvious reasons, including critical reliability, e.g. for medical diagnosis. Large language models have been called *stochastic parrots* in that they use probability to combine text that they

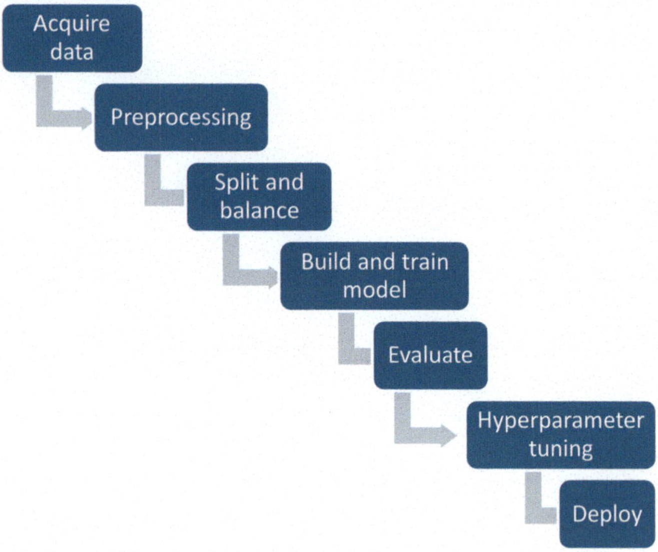

Fig. 3.11 Workflow for a deep learning project

have encountered before, but there may be more human-like reasoning qualities that are as yet unexplained (Hutson 2024). Casting light into the black boxes is an ongoing subject which itself could lead to progress in creating true artificial intelligence.

Workflows

Using a deep learning architecture in practice involves a workflow leading from data input to model implementation. The basic steps, adapted from Codecademy (2024) are shown in Fig. 3.11 and summarised in turn.

Acquire Data

Biopharmaceutical data could be text from publications and reports, readouts from experiments, such as gene expression levels, or other parameters. Web-based information can be *scraped* using a number of *Application Programming Interfaces* (APIs) for example those used to extract data from PubMed e.g. PubMed Central for Developers (2025).

Preprocessing

The aim of preprocessing is to clean the data (removing outliers, filling gaps etc) deal with different feature types and scale them using normalization or standardization techniques. This requires conversion to numbers, so categories, such as cell type, mRNA expression level etc. can be assigned a unique integer or converted to one-hot encodings consisting of a series of columns containing 1 or 0 representing the presence or absence of a feature respectively (see Rojo-Echeburúa 2024 for more details). Normalization of the data creates uniformity by avoiding the domination of features having larger values compared to the other features or variables, an undesirable state for reliable machine learning. Standardisation can be used for the same purpose by deriving a Z-score by subtracting the mean and dividing by the standard deviation the value of each feature. The resulting mean of 0 and standard deviation of 1 places the different feature categories on the same scale.

Text input can be split into individual words, letters, or subwords, all called *tokens* and assigned an integer value according to the number of times it appears in the text. *The* and *is* are obvious examples of frequent words and would be assigned a low number. The text is presented as a string of numbers that the machine learning model can read, with subwords used to help convey meaning in context: e.g. *drug discovery is conducted in several phases* where *drug, discovery phases* could be considered as subwords. For the algorithm to deal with large volumes of text in parallel, each token should be the same length, and if not, a *pad* token used to extend it, or truncation of long sequences to trim to the same length. Clearly this process is a trade-off between retaining semantic meaning and using the tokens in an optimal way by equalising. Examples of the above are shown in Lokare (2023).

Split and Balance

At this stage, pre-processed data are normally split into *training* and *validation* datasets to train and evaluate the model. The proportion of validation data will depend upon the total fed to the model, with more being used if the total is small to represent the true spread of data. If there are large differences in the distribution of categories between training and evaluation datasets, there will be a need for stratification to avoid over representation of minor classes in one of the two sets.

Some classes may appear much more than others resulting in a model that is heavily biased towards predicting those majority classes. The techniques of *undersampling* and *oversampling* are applied to the training set only. Undersampling is achieved by removing data from the dominant majority class, while oversampling involves duplicating data occurring in minority classes. Needless to say, this could be a subjective operation and require human judgment to check the validity of the model.

Build and Train Model

A deep neural network model will be built with a specified number of layers and hidden units within each layer. A *loss function* is applied to measure the difference between the output expected from the model and the actual output obtained. An *optimizer* helps to improve the model by adjusting its parameters so that the loss function value is minimized, examples being *Stochastic Gradient Descent* (SGD) and ADAM (see Sect. 3.2.4.4). The *learning rate* plays a central role in the optimization process by determining the magnitude of the steps used in SGD. A high learning rate enables the model to learn rapidly by taking larger strides, and a low learning rate ensures meticulous, but slower, progress. As with data splitting, a careful balance between the different options is required for each model.

Evaluate

The performance of the model on the validation will dictate how it will perform on new, unseen data. Measures of *prediction* and *recall* can be used to provide a numerical representation of the performance. Prediction is the number of true positive results divided by the number of all samples predicted to be positive, while recall is the number of true positive results divided by the number of all samples that should have been identified as positive. The F_1*-score* presents these precision and recall values symmetrically as a single value, 1.0 being perfect precision and recall and 0 for neither. Alternatively, a *confusion matrix* can be used to evaluate the model; a basic explanation is given in Ahmed (2023).

Hyperparameter Tuning

Hyperparameters are the values selected for the learning rates and optimization processes to control the performance of the deep learning model, not the input such as text or images. Normal practice involves running a smaller model and increasing hyperparameter sizes until training and performance start to diverge. Two issues *overfitting* and *underfitting*, could arise and must be corrected before the model is deployable. Overfitting is when the model fits too closely or exactly to its training data. As a result, the model cannot make accurate predictions or conclusions from any data other than the training data thus defeating the object of the exercise. Underfitting results from a model that is too simple, perhaps through needing longer training times or more input data. The model's performance can have a final evaluation before deployment by using a so-called *holdout set* of parameters that were retained from the validation set during data splitting. A good performance from the holdout set should give confidence that the model should work with unseen data as it is designed to do.

Deployment

Examples of deep learning foundation models have been given earlier. Some, like ChatGPT are made available to subscribers anywhere in the world. Biopharma researchers will more likely use proprietary models developed for in-house use or as a service for commercial use. Deployment of large models at scale requires the use of substantial computing resources (see later) and software must be available to provide an interface between the human end-user and the deep learning algorithms. *Containers* are software systems that package code and interfaces in one environment that can be readily deployed as required, one example being Docker (2024).

3.2.4.6 Emerging and Future Issues with AI/ML

AI/ML is a computational tool requiring connectivity between end users and the hardware running the models. There is no shortage of potential users: the number of people using the internet in 2024 is estimated to be 5.35 billion, or about 66% of the world's population rising to 7.9 billion in 2029 (Statista 2024). However, these are early days with AI and machine learning, particularly deep learning (DL). Examples of applications appear almost daily and there is a feeling that these technologies will dictate the future of life in general along with advances in drug discovery and development. However, as with any rapidly emerging field that is subject to some degree of hype, there are issues that need to be confronted, and obstacles overcome. This section starts with an analysis of the performance of DL and other machine learning models and the current state of progress in their ability to find novel drug targets and candidates.

Performance

Anyone who uses predictive text or chatbots will realise that some machine learning algorithms generally work, despite some frustrating errors that can be either amusing or annoying. Large language models like ChatGPT can produce remarkable results, see this book's Preface for example, but not always. The phenomenon of *hallucination*, in which the models produce nonsensical text is well known; a subset of these called *confabulations* by Farquar et al. (2024) could be overcome using the statistical methods these authors apply to the problem.

Another problem to be addressed is the danger of bias due to over- or under-emphasis of specific categories, like ethnicity, which has to be addressed by careful selection of input data categories. Perhaps a more fundamental problem with current deep machine learning algorithms is their inability to learn continuously unlike the human brain. The models go through a training phase and the selected weights frozen before deploying the system against new data. The models have to be rebuilt before incorporating large new data sets which is highly wasteful in resources. The ideal solution is to develop a system of continuous learning that adapts *on the fly*. Dohare et al. (2024) discuss this real problem with current deep learning and offer a potential way forward.

The applications of AI/ML in drug discovery will include natural language models and language translation, so the broader issues affecting these algorithms will apply here. Deep learning models like AlphaFold for protein structure prediction[6] are working very effectively and will no doubt be further refined over the years. One way of indirectly assessing performance is by gauging the degree of enthusiasm for AI/ML technology by pharma and biotech companies. A recent article in the Pharmaceutical Journal covers this well (Brazil 2024). Naturally all of the major pharma companies will have some investment in the field using both in-house expertise and collaborations with smaller biotechs. The latter are proliferating and include those listed in Table 3.3.

Actual physical progress being made in drug discovery can be measured by the number of AI-originated drug candidates in clinical trials. Jayatunga et al. (2024) have collated and analysed this information which as yet, does not include Phase III trials. Their analysis was based on >100 *native AI-based biotechs.*[7] 67 AI-based molecules (~50% in oncology) were in clinical trials in 2023, mostly Phase I. At the

[6] To be covered in Chap. 5

[7] These numbers are much greater than the representative examples shown in Table 3.3 and can be accessed from the Jayatunga et al. (2024) publication

Table 3.3 Some companies focused on AI/ML in drug discovery and development

Company	Activity	Location
Ainnocence	Protein engineering	Miami, USA
BenevolentAI	Drug discovery	London, Cambridge UK
Exscientia	Drug discovery	Oxford, UK
Generate Biomedicines	Protein engineering	Somerville, USA
Insilico Medicine	Drug discovery	Boston, USA
Lantern Pharma	Precision oncology	Dallas, USA
Recursion	Precision oncology & rare diseases	Salt Lake City, USA

end of 2023, 21 molecules had completed Phase I successfully, with only three failures, an impressive rate of success compared with conventional drug discovery. The 21 candidates were subdivided as follows:

3 repurposed molecules
9 AI-discovered targets
4 AI-designed small molecules
1 AI-designed vaccine
4 AI-designed antibodies

Of the ten molecules were tested in Phase II, four were successful, which is a similar proportion to conventional discovery. These figures are small, and some caveats have been highlighted by Jayatunga et al. (2024).

However, to reiterate the comments made at the beginning of this section, these are early days and there is clearly a great deal of promise in using AI/ML technology to produce drug candidates rapidly and efficiently, leading to an improved success rate in the clinic when compared with conventional approaches.

Feeding the Models

Foundation models like ChatGPT use vast amounts of training data *scraped* from various sources. While no reliable figure seems to exist for the size of training set used for ChatGPT4 for example, it would appear to use over one trillion parameters coded in terabytes of data. Given the extensive range of data sources used there are inevitable issues with copyright which are not fully resolved. For example, the *New York Times* are taking legal action against Microsoft and Open AI claiming that they are using their paper's content without permission. Other users of content can avoid this issue by purchasing licenses from originators, while scientific publishers are entering into commercial arrangements to allow use of their material. Not everyone is comfortable with this of course, but the ongoing and relentless acquisition of data seems inevitable. One day however, existing data for training general foundation models will run out. This situation has been reviewed by Jones (2024) who references the paper by Villalobos et al. (2024) estimating this to occur between 2026 and 2032.[8] This applies to existing data, but not to new material itself generated by AI/ML. There are dangers here: the worlds corpus of information could (and probably is) being polluted by LLM output, and the performance of deep neural networks fed such data is suspect (the AI model *collapses*). A striking example is provided by Shumailov et al. (2024) in which a model is initially trained on text relating to medieval architecture and the output fed back over several generations of running the model. The first few runs make sense, but model nine comes up with different coloured populations of jackrabbits, clearly nothing to do with architecture!

Proprietary data used for drug discovery/development will include that available in genomic databases, PubMed and many other sources and which are increasing all the time. As an example, *Recursion* state that they have >50 petabytes of molecular and clinical data available for training their models.

Intellectual Property

In Silicon Valley, we talk about software patents being the sport of kings, to quote Vijay Pande from Andreessen Horowitz (Hutson 2023). He is referring to the costs involved in filing and defending patents, so from a business perspective there will always be a case for patenting the drug molecules themselves but not necessarily the computational methods used to obtain them. However, given the potential advantages of AI in speeding up drug discovery, some patenting will be inevitable. According to a preliminary estimate by M Aboy, around 3000 patents have been filed to date on AI methods related to drug development and delivery (Hutson 2023). However, US patent law can see AI algorithms as abstract ideas and therefore not patentable. We can assume that the area will become clearer once drug candidates are successfully developed by AI and enter the marketplace.

Environmental Costs

The above sections have highlighted how AI/ML is and can make an impact on drug discovery and development as well as broader human activity. However, this comes at a huge cost in water and electricity consumption. For example, global AI demand in 2027 could be accountable for 4.2–6.6 billion cubic meters of water withdrawal, or half that of the United Kingdom (Li et al. 2023). To quote from Jafari et al. (2024) the US Energy Information Administration (EIA) projected that US data centres will consume about 88 terawatt-hours (TWh) annually by 2030, about 1.6 times the electricity consumption of New York City. The basic problem for projection of future power demands for generative AI is that nobody knows how events will unfold (Chen 2025). Energy efficiencies could be made, but if there is much greater uptake of the systems, power demand could rise even higher. Maybe nuclear fusion will provide cheap plentiful electricity by the end of the century, but decades will pass before it becomes viable, if at all.

[8] Villalobos estimates the stock of internet text tokens in 2024 to be 3100 trillion tokens

3.2.5 Hardware and Software

3.2.5.1 Big Data

Digital data are being generated and used at a staggering rate, so the term big data seems entirely appropriate.[9] To summarise some points made by Djuraskovic (2024):

Around 79 zettabytes of data[10] estimated to be produced in 2021, doubling in 2025.
Approximately 90% of the data are replicated, 10% being novel.
It would take approximately 181 million years to download all data from the web today.
The market for healthcare big data analytics could reach $67.82 billion by 2025.

Unsurprisingly, the concept of big data has been embraced by biopharma companies who use disparate sources of information to make each stage of the discovery and development pipeline more efficient. A McKinsey report (Cattell et al. 2013) clearly laid out the possibilities over 10 years ago, including the following uses of big data:

Increasing the efficiency of drug and drug target discovery.
Optimising the selection of clinical trial participants using genetic and other data and recruitment though analysing social media in addition to conventional medical channels.
Seamlessly integrating different departments and functions within a highly distributed and complex organisation.

3.2.5.2 Compute

Processing

The hardware and software resources required to implement machine learning and other algorithms are compute resources or just *compute*. The physical components required include memory devices (RAM), graphics processors, and in conventional computer architectures, central processing units (CPUs). The last components consist of arrays of transistors switching from 0 s and 1 s to represent binary code. Individual transistors in arrays can now be miniaturised to the nanometer scale and be packed at a density of >100 million per mm^2 (Garisto 2024). However, there are fundamental constraints in building chips able to perform tasks with the massive datasets used in deep learning, for example. Graphical Processing Units (GPUs) represent an alternative architecture where thousands of separate processing *cores* are used in parallel compared with the small number used in CPUs. The increase in speed afforded by GPUs enables processing of deep learning algorithms in a fraction of the time that would be taken by CPUs, but not without financial implications.[11] At

[9] Data are plural!

[10] One zettabyte is 10^{21} bytes or one sextillion

[11] GPT3 has 175Bn parameters run on 1.024 GPUs for 1 month and costing several million dollars (Garisto 2024)

the time of writing, Nvidia is the leader in providing GPUs for deep learning, but other architectures are under development and may come to the fore in the future (Garisto 2024).

Neuromorphic Computing

Neuromorphic computers are based on the structure of the brain and consist of neurons and synapses that are integrated to store memory and process data. A detailed survey of the state of neuromorphic computing and its future implementation has been published recently Kudithipudi et al. (2025).

Conventional von Neumann architectures separate CPUs from memory units, the latter encoding numerical information as binary digits Neuromorphic computers are based more on neurobiology in that information is received as input spikes with their timings, magnitude and shape used to encode numerical information (Schuman et al. 2022).

These are early days for creating neuromorphic computers capable of outperforming conventional hardware but significantly, the systems consume much less energy. This has obvious implications for training and operating deep learning models as well as miniaturising hardware for embedding in phones and other devices. Lastly, research is ongoing to move from silicon-based neuromorphic computers to those based on ferroelectric, optical or other physical properties.

Cloud Computing

The availability of cheap computer hardware has made *computer,* or *server farms* an attractive economic proposition for organisations as well as being useful for crowdsourcing campaigns such as encouraging amateurs to classify astronomical images. Large numbers of individual PCs can be linked together to provide significant parallel processing power. However, given the data storage and processing demands of many aspects of digital life, including deep learning, there is a growing need for distributed computing in the form of *The Cloud.* According to Statista (Vailshery 2024) the global public cloud computing market is expected to reach an estimated $825Bn in 2025. Cloud computing uses the internet to connect users requesting computer services to a platform provided by vendors such as Microsoft, Amazon, Alibaba or Google. A central server interfaces between the client and cloud servers and handles compute projects on an *ad hoc* basis. The compute demands of deep learning can be met with services like Oracle's Cloud Infrastructure (OCI) Supercluster which has up to 32,768 graphics processing units (GPUs) with the future potential of 65,536 (OCI 2024).

Quantum Computing

In the future, conventional computing using binary code may find serious competition from *quantum computers*, a high-profile technology in early development. Quantum computing uses quantum bits or *qubits* based on the coherent superposition of ones and zeros; unlike conventional bits having a single value at any one time. a qubit can be both zero and one at the same time and in different proportions. This makes it possible for a quantum computer with just 30 qubits to perform

10 billion floating-point operations per second, about 5.8 billion more than the most powerful PlayStation video game console on the market (Iberdrola 2024). The attractions of this enormous speed are not lost on biopharmaceutical companies, with major players forming collaborations and developing IP (Chandran 2024). The development of quantum computing development in general is seen as a high priority for the future (along with AI/ML) and as an example, the Novo Nordisk Foundation is investing €188 M in a quantum technology ecosystem that will also contribute to R&D in the life sciences (Novo Holdings 2024).

While the attractions of quantum computing are clear, there are serious technical obstacles to be overcome. The main problem lies in the process of *decoherence* in which the superposed qubits are readily disrupted by external stimuli, like radiation. A major challenge in producing quantum computers lies in shielding the superposed qubits for enough time to perform calculations. Quantum error correction is a key component and is mathematically different to classical error correction of the type found in digital media.[12] Physical components must be made from materials that minimise decoherence and the computer cooled to hundredths of a degree above absolute zero using liquid helium. Given the progress to date and the intense activity in this area, it seems highly likely that quantum computing, along with AI and machine learning, will have a major impact on life in general including drug discovery and development.

References

3Blue1Brown (2016) But what is a neural network? https://www.youtube.com/watch?v=aircAruvnKk&list=PLZHQObOWTQDNU6R1_67000Dx_ZCJB-3pi. Accessed 20 Mar 2025

Acharjee A et al (2016) Integration of metabolomics, lipidomics and clinical data using a machine learning method. BMC Bioinformatics 17:440. https://doi.org/10.1186/s12859-016-1292-2. s

Ahmed NA (2023) What is a confusion matrix in machine learning? The model evaluation tool explained. Datacamp https://www.datacamp.com/tutorial/what-is-a-confusion-matrix-in-machine-learning. Accessed 20 Mar 2025

AWS (2025) AWS: what are transformers in artificial intelligence? https://aws.amazon.com/what-is/transformers-in-artificial-intelligence/. Accessed 20 Mar 2025

Bennett JM, Kendrew JC (1952) The computation of fourier synthesis with a digital electronic calculating machine. Acta Crystallogr 5:109–117. https://doi.org/10.1107/S0365110X52000228. s

Biological Expression Language (2024). https://biological-expression-language.github.io/. Accessed 20 Mar 2025

Biorelate (2025). https://biorelate.com/company. Accessed 20 Mar 2025

Brazil R (2024) How AI is transforming drug discovery. The Pharmaceutical Journal. https://pharmaceutical-journal.com/article/feature/how-ai-is-transforming-drug-discovery. Accessed 20 Mar 2025

Brooks M (2024) The maths that drive AI. Nature 631:244–246. https://doi.org/10.1038/d41586-024-02185-z. Accessed 20 Mar 2025

Brown TB et al (2020) Language models are few-shot learners. Adv Neural Inf Proces Syst 33:1877–1901. https://doi.org/10.48550/arXiv.2005.14165. Accessed 20 Mar 2025

[12]For example, over 50% of the information on an audio CD is dedicated to error correction

Brownlee GG (2013) Frederick Sanger (1918–2013). Curr Biol 23:PR1074–PR1076. https://doi.org/10.1016/j.cub.2013.11.037. Accessed 20 Mar 2025

Butcher EC et al (2004) Systems biology in drug discovery. Nat Biotech 22:1253–1259. https://doi.org/10.1038/nbt1017. Accessed 20 Mar 2025

Buttenschoen M et al (2024) PoseBusters: AI-based docking methods fail to generate physically valid poses or generalise to novel sequences. Chem Sci 15:3130–3139. https://doi.org/10.1039/D3SC04185A. Accessed 20 Mar 2025

Castelvecchi D (2024) Researchers built an 'AI scientist' – what can it do? Nature 633:266. https://doi.org/10.1038/d41586-024-02842-3. Accessed 20 Mar 2025

Cattell J et al (2013) How big data can revolutionize pharmaceutical R&D. https://www.mckinsey.com/~/media/McKinsey/Industries/Pharmaceuticals%20and%20Medical%20Products/Our%20Insights/How%20big%20data%20can%20revolutionize%20pharmaceutical%20R%20and%20D/How%20big%20data%20can%20revolutionize%20pharmaceutical%20RD.pdf. Accessed 20 Mar 2025

Centre for Computing History (2025). https://www.computinghistory.org.uk/cgi/computing-timeline.pl. Accessed 20 Mar 2025

Chandran A (2024) Biopharma foresees a 'quantum advantage': they could be right. Nat Biotech 42:687–693. https://doi.org/10.1038/s41587-024-02251-x. Accessed 20 Mar 2025

Chemoinformatics in Drug Discovery Oprea TI ed. (2005) Wiley-VCH Verlag. https://doi.org/10.1002/3527603743. Accessed 20 Mar 2025

Chen S (2025) How much energy will AI really consume? The good, the bad and the unknown. Nature 639:22–24. https://doi.org/10.1038/d41586-025-00616-z. Accessed 20 Mar 2025

Codecademy (2024) Deep learning workflow. https://www.codecademy.com/article/deep-learning-workflow. Accessed 20 Mar 2025

Corso G et al (2023) DiffDock: diffusion steps, twists, and turns for molecular docking. arXiv:2210.01776v2. Accessed 20 Mar 2025

Dara S et al (2022) Machine learning in drug discovery: a review. Artif Intell Rev 55:1947–1999. https://doi.org/10.1007/s10462-021-10058-4. Accessed 20 Mar 2025

de Chadarevian S (2018) John Kendrew and myoglobin: protein structure determination in the 1950s. Protein Sci 27:1136–1143. https://doi.org/10.1002/pro.3417. Accessed 20 Mar 2025

Djuraskovic O (2024) Big data statistics 2023: how much data is in the world? https://firstsiteguide.com/big-data-stats/. Accessed 20 Mar 2025

Docker (2024). https://www.docker.com/why-docker/. Accessed 20 Mar 2025

Dohare S et al (2024) Loss of plasticity in deep continual learning. Nature 632:768–774. https://doi.org/10.1038/s41586-024-07711-7. Accessed 20 Mar 2025

Dörpinghaus J et al (2022) Context mining and graph queries on giant biomedical knowledge graphs. Knowl Inf Syst 64:239–1262. https://doi.org/10.1007/s10115-022-01668-7. Accessed 20 Mar 2025

EPA SMILES Tutorial (2025). https://archive.epa.gov/med/med_archive_03/web/html/smiles.html. Accessed 20 Mar 2025

Farquar S et al (2024) Detecting hallucinations in large language models using semantic entropy. Nature 630:625–630. https://doi.org/10.1038/s41586-024-07421-0. Accessed 20 Mar 2025

Garisto D (2024) How cutting-edge computer chips are speeding up the AI revolution. Nature 630:544–546. https://doi.org/10.1038/d41586-024-01544-0. Accessed 20 Mar 2025

Gauthier J et al (2019) A brief history of bioinformatics. Brief Bioinform 20:1981–1996. https://doi.org/10.1093/bib/bby063. Accessed 20 Mar 2025

Ghasemi A et al (2023) Scientific publishing in biomedicine: a brief history of Scientific Journals. Int J Endocrinol Metab 21:e131812. https://doi.org/10.5812/ijem-131812. Accessed 20 Mar 2025

Good A (2007) Virtual screening. In: Comprehensive medicinal chemistry II, vol 4. Elsevier, pp 459–494. https://doi.org/10.1016/B0-08-045044-X/00262-5. Accessed 20 Mar 2025

H2O.ai Wiki (2025). https://h2o.ai/wiki/. Accessed 20 Mar 2025

Hutson M (2023) AI for drug discovery is booming, but who owns the patents? Nat Biotech 41: 1491–1503. https://doi.org/10.1038/s41587-023-02029-7. Accessed 20 Mar 2025

Hutson M (2024) How does ChatGPT think? Nature 629:986–988. https://doi.org/10.1038/d41586-024-01314-y. Accessed 20 Mar 2025

Iberdrola (2024) All about quantum computing. https://www.iberdrola.com/innovation/what-is-quantum-computing. Accessed 20 Mar 2025

IBM Research (2022). https://research.ibm.com/blog/what-are-foundation-models. Accessed 20 Mar 2025

IBM Think: what are convolutional neural networks? (2025). https://www.ibm.com/topics/convolutional-neural-networks. Accessed 20 Mar 2025

Jafari T et al (2024) Projecting the electricity demand growth of generative AI large language models in the US. Center on Global Energy Policy. https://www.energypolicy.columbia.edu/projecting-the-electricity-demand-growth-of-generative-ai-large-language-models-in-the-us/#. Accessed 20 Mar 2025

Jain V (2023) How AI could lead to a better understanding of the brain. Nature 623:247–250. https://doi.org/10.1038/d41586-023-03426-3. Accessed 20 Mar 2025

Jayatunga MKP et al (2024) How successful are AI-discovered drugs in clinical trials? A first analysis and emerging lessons. Drug Discov Today 29:1–6. https://doi.org/10.1016/j.drudis.2024.104009. Accessed 20 Mar 2025

Jones N (2024) The AI revolution is running out of data. What can researchers do? Nature 636:290–292. https://doi.org/10.1038/d41586-024-03990-2. Accessed 20 Mar 2025

Kudithipudi D et al (2025) Neuromorphic computing at scale. Nature 637:801–812. https://doi.org/10.1038/s41586-024-08253-8. Accessed 20 Mar 2025

Levitt M (2001) The birth of computational structural biology. Nature 8:392–393. https://doi.org/10.1038/87545. Accessed 20 Mar 2025

Li Y et al (2018) Multi-objective de novo drug design with conditional graph generative model. J Cheminform 10:33. https://doi.org/10.1186/s13321-018-0287-6. Accessed 20 Mar 2025

Li P et al (2023) Making AI less "Thirsty": uncovering and addressing the secret water footprint of AI models. https://doi.org/10.48550/arXiv.2304.03271. Accessed 20 Mar 2025

Lokare G (2023) Preparing text data for transformers: tokenization, mapping and padding. https://medium.com/@lokaregns/preparing-text-data-for-transformers-tokenization-mapping-and-padding-9fbfbce28028. Accessed 20 Mar 2025

Maia EHB et al (2020) Structure-based virtual screening: from classical to artificial intelligence. Front Chem 8. https://doi.org/10.3389/fchem.2020.00343

Moor J (2006) The Dartmouth College artificial intelligence conference: the next fifty years. AI Mag 27:87. https://doi.org/10.1609/aimag.v27i4.1911. Accessed 20 Mar 2025

Narkhede S (2018) Understanding AUC – ROC curve. Towards data science. https://towardsdatascience.com/understanding-auc-roc-curve-68b2303cc9c5. Accessed 20 Mar 2025

Nicholson DN, Greene CS (2020) Constructing knowledge graphs and their biomedical applications. Comput Struct Biotechnol J 18:1414–1428. https://doi.org/10.1016/j.csbj.2020.05.017. Accessed 20 Mar 2025

Novo Holdings (2024). https://novoholdings.dk/news/novo-holdings-commits-dkk-1-4-billion-euro188-million-to-build-a-quantum-technology-start-up-ecosystem. Accessed 20 Mar 2025

OCI (2024) AI infrastructure. https://www.oracle.com/ai-infrastructure/. Accessed 20 Mar 2025

Offord C (2024) 'Google for DNA' indexes 10% of world's known sequence data. Science 384: 1053–1054. https://doi.org/10.1126/science.zam1hsh. Accessed 20 Mar 2025

Open AI (2024). https://openai.com/research/. Accessed 20 Mar 2025

PubChem (2025). https://pubchem.ncbi.nlm.nih.gov/compound/Morphine. Accessed 20 Mar 2025

PubMed Central for Developers (2025). https://www.ncbi.nlm.nih.gov/pmc/tools/developers/. Accessed 20 Mar 2025

Pun FW et al (2023) AI-powered therapeutic target discovery. Trends Pharmacol Sci 44:561–572. https://doi.org/10.1016/j.tips.2023.06.010. Accessed 20 Mar 2025

Rojo-Echeburúa A (2024) What is one hot encoding and how to implement it in Python. DataCamp. https://www.datacamp.com/tutorial/one-hot-encoding-python-tutorial?dc_referrer=https%3A%2F%2Fwww.google.com%2F. Accessed 20 Mar 2025

Sadybekov AV, Katritch V (2023) Computational approaches streamlining drug discovery. Nature 616:673–685. https://doi.org/10.1038/s41586-023-05905-z. Accessed 20 Mar 2025

Savage N (2023) Drug discovery companies are customizing ChatGPT: here's how. Nat Biotech 41:585–596. https://doi.org/10.1038/s41587-023-01788-7. Accessed 20 Mar 2025

Schuman CD et al (2022) Opportunities for neuromorphic computing algorithms and applications. Nat Comput Sci 2:10–19. https://doi.org/10.1038/s43588-021-00184-y. Accessed 20 Mar 2025

Shumailov I et al (2024) AI models collapse when trained on recursively generated data. Nature 631:755–758. https://doi.org/10.1038/s41586-024-07566-y. Accessed 20 Mar 2025

Stanford University AI Index Report (2024). https://aiindex.stanford.edu/report/. Accessed 20 Mar 2025

Statista (2024). https://www.statista.com/statistics/617136/digital-population-worldwide/. Accessed 20 Mar 2025

Stent GS (1968) That was the molecular biology that was. Science 160:390–395. https://doi.org/10.1126/science.160.3826.390. Accessed 20 Mar 2025

UK Biobank (2025). https://www.ukbiobank.ac.uk/learn-more-about-uk-biobank/about-us. Accessed 20 Mar 2025

Uspenskyi S (2024) How many software engineers are there in 2024? https://springsapps.com/knowledge/how-many-software-engineers-are-there-in-2024. Accessed 20 Mar 2025

Vailshery LS (2024) Public cloud services market size 2017–2024. Statista. https://www.statista.com/statistics/273818/global-revenue-generated-with-cloud-computing-since-2009/#statisticContainer. Accessed 20 Mar 2025

Villalobos P et al (2024) Will we run out of data? Limits of LLM scaling based on human-generated data. https://doi.org/10.48550/arXiv.2211.04325. Accessed 20 Mar 2025

Voit EO (2022) Perspective: systems biology beyond biology. Front Syst Biol 2. https://doi.org/10.3389/fsysb.2022.987135. Accessed 20 Mar 2025

Weka.io AI Guide in Simple Terms (2025). https://www.weka.io/learn/ai-ml/what-is-ai/. Accessed 20 Mar 2025

Wellcome Trust Report (2023) Unlocking the potential of AI in drug discovery. https://wellcome.org/reports/unlocking-potential-ai-drug-discovery. Accessed 20 Mar 2025

Part II

Emerging Opportunities

Points of Scientific Impact 4

4.1 Introduction

Part I covered the state of drug discovery and development at end of the first quarter of the twenty-first century. The focus of Part II is to highlight the expanding range of possibilities for discovering and developing medicines to treat well-known diseases and to anticipate those that are yet to appear, such as new viral infections. The discovery of new drugs must also occur in parallel with preventative medicine and developing the diagnostics required to catch diseases at a stage when they have a good chance of being cured or at least delayed. What the next 75 years will bring is obviously a matter of conjecture, with many factors having a profound impact on humanity other than biopharma R&D.[1] However, it does no harm to think of the features in an ideal medicine that are, or should be, the focus of current and future drug discovery efforts. Such features could be summarised as follows:

Orally bioavailable molecules that Induce disease remission, or at least long-term control in a wide patient population, are well tolerated and cost effective.

Oral bioavailability has long been seen as ideal because of the ease with which drugs can be administered and manufactured. Unfortunately this is often unattainable where the drug target is undruggable with small molecules or is part of a complex set of interacting systems as in tissue regeneration, or the immune and nervous systems. The result has been the development of large molecules, ranging from proteins, including antibodies, to cell and gene therapies. While these can be curative, they are expensive to produce and may only be effective in a small patient population. A future objective therefore would be to replace biotherapeutics with small molecules that achieve the same result. Alternatively, where this is impossible, new approaches for the oral delivery of large molecules could be a viable option.

[1] Of course the world is complicated, and many other aspects of human existence are in play which for obvious reasons are not within the scope of this book

E. D. Zanders, *Twenty-first Century Drug Discovery*,
https://doi.org/10.1007/978-3-031-93949-5_4

Alterations at the genomic level though gene editing, while very expensive initially, could be economically viable if the disease is halted before the need for a lifetime of expensive treatments. The same applies to cell therapy, but for potentially curative common diseases these will have to be made available *off the shelf* rather than manufactured individually.

4.2 The Drug Discovery Pipeline Revisited

The drug discovery pipeline was introduced in Chap. 2. and is revisited here in Fig. 4.1 to form a point of reference for the scientific developments covered in the remaining chapters of this book.

The main disciplines covering each part of the pipeline are shown and may, like genomics, be involved at multiple stages. Machine learning (as one of the main components of AI) has been placed separately to illustrate its potential impact at all stages.

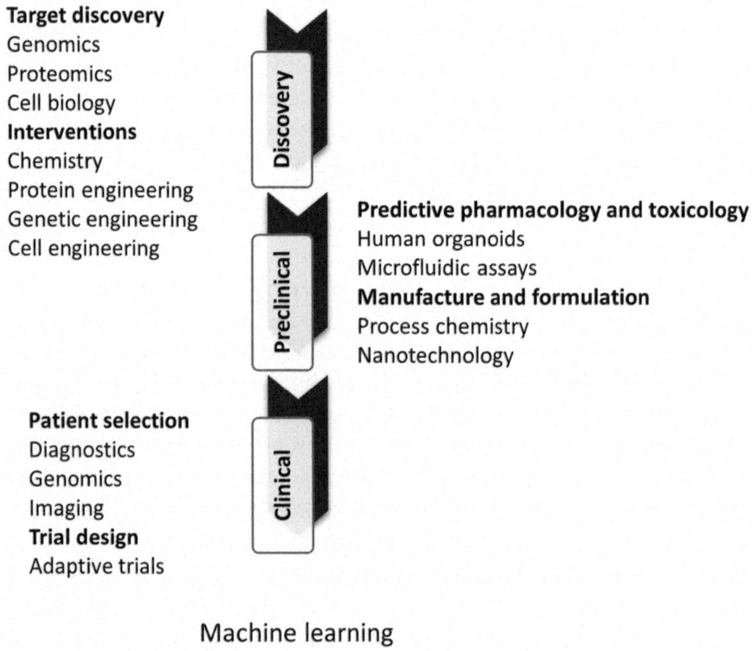

Fig. 4.1 Annotated drug discovery pipeline

4.3 Systems in Biopharmaceutical R&D

The field of drug discovery and development continues its relentless expansion and shows no sign of abating. This makes it harder than ever to cover the relevant material in a clear and comprehensible way.[2] One way of creating clarity is to identify the different scientific areas used in drug discovery and development and organise them into coherent groups. An example of this is shown in Fig. 4.2 where four groups are presented, each with covering a *system* in the broadest sense. Naturally these are not mutually exclusive categories since each will impact upon the other in various ways. The microbiome, immune and nervous systems are chosen because of their significant involvement in health and disease and therefore in drug discovery. Although these have already been listed in Chap. 2. when covering target diseases, the aim of the following section is to provide brief summaries of the systems listed in Fig. 4.2, with later chapters providing more in-depth detail where appropriate.

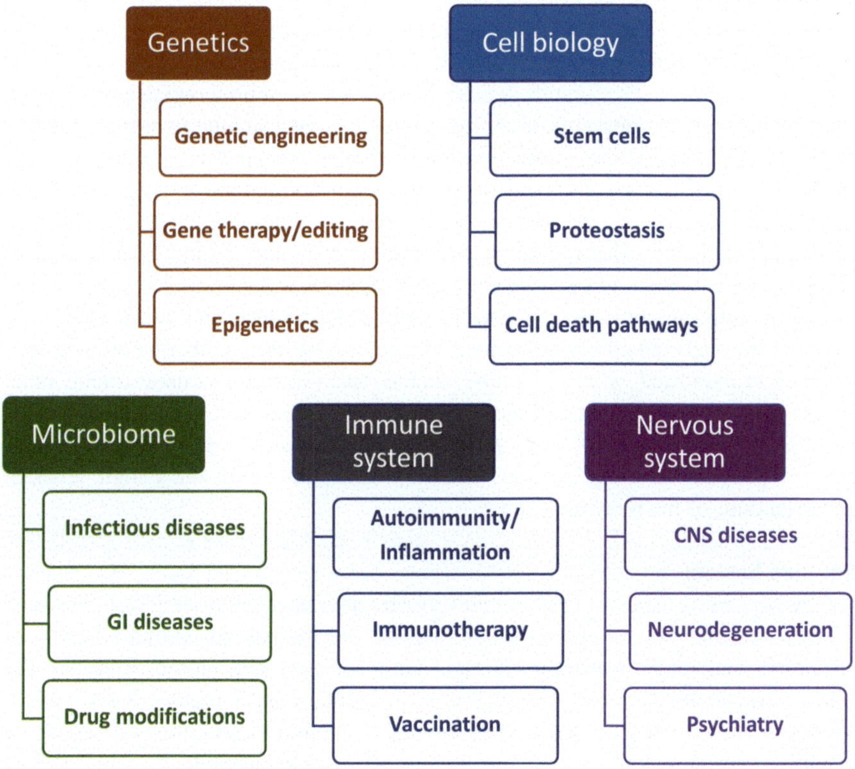

Fig. 4.2 *Systems* in biopharma R&D

[2]The main incentive of course for writing this book

Genetics

Although formally the study of inheritance, genetics in the context of Fig. 4.2 encompasses, genomics, proteomics, metabolomics and related fields. Central to biopharma R&D is the ability to manipulate nucleic acids and proteins for target discovery and drug development using the tools of genetic engineering and gene therapy/editing. Epigenetics/epigenomics underpins the regulation of gene expression in health and disease and presents exciting opportunities for novel drug discovery.

Cell Biology

Cells can be used as therapies in the form of stem cells or engineered to create new functions, for example tumour cell killing. However, there are many opportunities for target discovery connected with, among other things, protein quality control (*proteostasis*), or the various forms of cell death including *apoptosis*.

Microbiome

The term *microbiome* describes the commensal microbes and viruses associated with the body but is broadened here to include external agents that are causative of infectious diseases. The search for drug targets and antimicrobials/antivirals is long established and ongoing, aided by genomics and machine learning. Another aspect of the microbiome is the interplay between commensal bacteria and the immune and nervous systems which provides opportunities for treating inflammatory bowel diseases and even neurological conditions. Microbes have a documented effect on certain chemotherapy drugs, which adds a new and actionable dimension to cancer treatment. Microbes can be used as drugs in their own right, as probiotics to create a healthy microbiome and remove pathogenic bacteria and to introduce new functions through genetic engineering. The role of viruses as therapeutics is well known and exploited as gene therapy vectors and oncolytic viruses among other things. Lastly, and by no means least, vaccines designed to tune the immune system towards sustainable protective immunity depend upon choosing appropriate microbial components, delivery systems and adjuvants, all based on a high level of understanding of the relevant science.

Immune System

The two opposing aspects of the immune system are important in drug discovery and development namely stimulating immunity for vaccination and immunotherapy or selectively reducing its manifestation in autoimmunity and atopic diseases. The complexity of the immune system seems to increase on a regular basis as new soluble factors (cytokines etc) and subsets of lymphoid and myeloid cells are identified. The success of engineered cytotoxic T-cells in cancer therapy has opened the way for the development of alternative cell hosts including natural killer (NK) cells and monocytes. The fine details of transcriptional regulation and metabolic pathways in effector cells are being revealed that could lead to strategies to overcome resistance to tumour killing, or other scenarios. Lastly, the importance of the innate immune response is being recognised not only as a mediator of inflammation but also as an adjuvant for vaccination.

Nervous System

The central and peripheral nervous systems are complex but important targets for drug discovery and development. As laid out in the figure, CNS diseases cover conditions such as chronic pain and epilepsy as well as psychiatric diseases such as depression and anxiety. Neurodegeneration in the form of Alzheimer's, Parkinsons and Huntingdon's Diseases among others, is a separate but important area of chronic disease. The complex signalling pathways and multiple subtypes of serotonin, noradrenaline and dopamine receptors, for example, make pharmacological intervention particularly challenging. Activation or inhibition of the wrong receptor subtype or signalling pathway can lead to adverse drug reactions, including hallucination. However, this latter effect could be exploited through the devlopment of novel antidepressants based on hallucinogenic drugs modified to retain a therapeutic but not hallucinogenic effect. Ion channels are another target class with proven clinical success but again, require highly selective compounds to discriminate between closely related members. Recent success with antagonists of the NaV1.8 sodium channel is leading the way to a totally novel class of analgesics without the liabilitites of opiates.

Hunting for Targets

5

5.1 Introduction

The likely global burden of disease in the coming decades was covered in Chap. 1, with cancer the top of the list. It should come as no surprise therefore that a large amount of biopharma R&D is devoted to oncology. Of course, other disease areas vie for attention, including cardiovascular, respiratory, neurological and metabolic as highlighted in Chap. 2. Add rare diseases to the list and there are plenty of opportunities in drug development to keep drug discovery practitioners busy for many years.

The starting point for drug development is the identification of a suitable target using the strategies already outlined in Chap. 2. The current chapter provides an in-depth discussion of drug target discovery supported by a comprehensive series of primary references and reviews. The aim is to describe current research that has the potential to create the successful medicines of the future. Some of the most promising technologies include high-throughput protein sequencing, single cell analysis and machine learning.

It is relatively straightforward to identify potential targets among intracellular, extracellular or secreted molecules, but finding an intervention that translates into the clinic is far from assured. However, successes are achieved, and the more information that can be gathered about a target the better the decisions that can be made supporting its viability in a drug discovery programme.

This chapter highlights the main technical approaches[1] used for target identification now and for the foreseeable future. While aiming to focus on target discovery as much as possible, there will be some inevitable overlap with the different drug molecules found to interact with them, which is covered more specifically in Chap. 6, *Interventions*. The remainder of the chapter is devoted to examples of

[1] Including the 'systems' outlined in Chap. 4

E. D. Zanders, *Twenty-first Century Drug Discovery*,
https://doi.org/10.1007/978-3-031-93949-5_5

how the various technologies are being applied in practice and importantly, what the field of target discovery might look like in the future.

5.1.1 Historical Background

The following references have provided material for the historical sections below: Halliwell (2007), Prüll et al. (2009) and Zanders (2016).

Magic Bullets

In 1906, Paul Ehrlich included the following as part of an address to inaugurate the Georg-Speyer Haus in Frankfurt, Germany:

If we picture an organism as infected by a certain species of bacterium, it will obviously be easy to effect a cure if substances have been discovered which have an exclusive affinity for these bacteria and act deleteriously or lethally on these alone, while at the same time they possess no affinity for the normal constituents of the body and can therefore have the least harmful, or other, effect on that body. Such substances would then be able to exert their full action exclusively on the parasite harboured within the organism and would represent, so to speak, magic bullets, which seek their target of their own accord.

Receptors

This early concept of a drug target represented a transition point between the beginnings of the modern pharmacology that emerged in the nineteenth century and the description and eventual isolation of defined receptors for synthetic drug molecules that occurred in the twentieth. It is receptors, along with other small molecule targets such as enzymes, transporters and ion channels that form the basis of most drug discovery efforts. The receptor theory of drug action was not universally accepted until as late as the 1940s by which time pharmacology was developing into a mature, quantitative discipline. For example, concepts such as agonists, affinity and drug efficacy were well known by the 1950s and 60s, but little of this knowledge had been applied to pharmaceutical discovery. This all changed with the identification and exploitation of adrenaline receptor subtypes by Raymond Alquist and Sir James Black respectively. In 1948, the American pharmacologist Alquist published a study in which he determined the rank order of potency of a series of compounds designed to relax uterine muscles in cases of dysmenorrhoea. The compounds, including adrenaline (epinephrine in the USA) excited or inhibited various tissues and revealed the presence of two classes of adrenergic receptors alpha and beta. Alquist considered the idea of receptors as a theoretical tool and the later subdivisions of adrenoreceptors into α_1, α_2, β_1 and β_2 subtypes caused him anxiety (β_3 came much later). He believed that *if there are too many receptors, something is obviously wrong*. What he would have made of our current inventory of receptors and subtypes is probably best left to the imagination. Later on, Sir James Black at the UK company Imperial Chemical Industries (now part of AstraZeneca) identified beta adrenergic receptor antagonists (beta blockers) and later, antagonists of histamine receptor subtypes, both leading to significant support for pharmacology and medicinal chemistry and a rapid expansion of the pharmaceutical industry through sales of the resulting blockbuster drugs.

By the 1960s the receptor theory of drug action was accepted by pharmacologists, but still not understood at the molecular level. Early work on drug receptors provided hints that they were located in specialised regions of tissues. R.P. Cook showed in 1926 that acetylcholine action on frog muscle was blocked by methylene blue dye before it stained the muscle tissue. This antagonist action was reversible, as demonstrated by washing away and reapplying the methylene blue even though the heart muscle retained the blue staining throughout. This suggested that methylene blue had reversibly bound to receptors located at the cell surface. Much later in the late 1960s, Eduardo Robertis and colleagues disrupted tissue with detergents and isolated synaptosomes by differential centrifugation. Synaptosomal membranes contain the nicotinic acetylcholine receptors that were the first receptor molecules to be purified, later achieved in the early 1970s by Jean-Pierre Changeux and Ricardo Miledi using the electric organ of rays and eels as rich sources of receptor protein. This period saw the introduction of radioligand binding and affinity purification with potent ligands, in this case α-bungarotoxin. The receptor was then shown to be a 275KDa complex formed from multiple protein subunits and is now known to be one of some 300 ion channels, many of which are of major pharmaceutical interest.

Another class of receptors, including those binding adrenaline and histamine, were identified from the protein sequence of bovine rhodopsin. This protein, involved in the visual system, has a serpentine structure traversing the cell membrane seven times and associating with a G-protein complex for signal transduction. Accordingly, this receptor class is identified as GPCRs (G-protein coupled receptors) 7TM receptors (seven transmembrane) or serpentine receptors.

Enzymes

The true nature of enzymes as catalytic proteins was not known until Sumner's crystallisation of urease in 1926 and Northrop's studies on pepsin in 1929. Nevertheless, enzyme inhibition was understood at this time and was exploited, for example, in the development in the 1930s of cholinesterase inhibitors that could be used in glaucoma treatment; unfortunately they had more potential as nerve agents for military use. Later in the 1940s the antibacterial sulphonamide drugs, by now being superseded by penicillin, provided a lead for novel diuretic drugs (thiazides) based on the inhibition of carbonic anhydrase in the renal tubule of the kidney. The list of enzyme targets discovered through the remainder of the twentieth century includes those for highly successful drugs used in the treatment of millions of patients; these include angiotensin converting enzyme for hypertension, cyclooxygenases as targets for anti-inflammatory drugs, HMGCoA reductase inhibitors for hypercholesterolaemia and tyrosine kinase inhibitors for oncology.

Early Genetic Engineering

The first molecular cloning experiment, in which ribosomal RNA from *Xenopus laevis* was transferred to *E.coli* in a plasmid vector, was published in 1974 (Morrow et al. 1974). Ten years later the first pharmacological receptor molecules were cloned (the α, β, γ and δ subunits of the nicotinic acetylcholine receptor from *T. californica* (historical review in Costanzi et al. 2009). This achievement was possible because sufficient amino acid sequence of the receptor protein was available to allow investigators to design degenerate oligonucleotides for screening cDNA libraries. This approach has been used many times since then to clone genes encoding a wide variety of human proteins, some of which are known drug targets. However, the only way to identify the full repertoire of human protein-encoding genes was to sequence the entire 3.5 gigabase genome, the results of which form the basis for much of what follows.

5.2 Current Approaches to Drug Target Discovery

5.2.1 Known Drugs and Unknown Targets

One of the points to come out of the early receptor studies outlined in the previous section was that effective drugs acting on these entities were available before any real idea of what the receptors actually were. This situation has arisen for many medicines that may have used since antiquity but without a known target(s), not something that modern drug discovery scientists feel comfortable with. Advances in cell and molecular biology plus sensitive analytical techniques have identified targets for, among others, natural products such as opiates, salicylates and the immunosuppressants cyclosporine A, FK506 and rapamycin. The opiates bind three classes of GPCR, MOP, KOP and DOP (formerly μ, κ and δ), the salicylates inhibit prostaglandin synthase (cyclooxygenases) and the immunosuppressants bind to immunophilins to affect cell signalling.

Traditional Medicines

Natural products from plants, animals and microorganisms have provided novel chemical entities as drugs plus targets that can be explored further with synthetic molecules tailored to a specific situation. Three of these product classes were mentioned in the previous section, but of course there are many more natural products with proven or potential use as therapeutics, not least traditional medicines. Pan et al. (2013) have written a comprehensive review of traditional medicines and how they can be aligned with modern thinking in the biopharma industry. Chinese herbal medicine (CHM) and Indian herbal medicine (IHM), influence modern healthcare with herbal medicines providing primary healthcare for approximately 3.5–4 billion people worldwide. It has been estimated that approximately 420,000 plant species exist, but very limited knowledge is available for most of them. Opportunities exist therefore for identifying new treatments ant drug targets while accepting some of the caveats discussed by Pan et al. (2013).

5.2.2 How Many Targets?

In considering future target opportunities in biopharma R&D it is useful to take stock of the number that have been identified for existing medicines, to build on existing discovery strategies, and to devise new ones.

5.2.2.1 Targets for Existing Drugs

Various attempts have been made over the years to catalogue the targets for approved drugs with Drews and Ryser (1997) among the first. They considered targets listed in Goodman and Gilman's, *The Pharmacological Basis of Therapeutics*, identifying 483 human and microbial entries. With the availability of the human genome sequence in the early twenty-first century, and various databases such as DrugBank from the University of Alberta in Canada, it was then possible to further refine the figures to 435 targets in the human genome (Rask-Anderson et al. 2011). More recently, Santos et al. (2017) published a *comprehensive map of molecular drug targets* curating 893 human and pathogen-derived biomolecules through which 1578 FDA-approved drugs act. A figure of 742 targets was published in 2016 as part of a wider survey into *potential* drug targets (Zanders 2016). Inevitably the number of targets for known drugs increases each year as new discoveries are approved by the regulators. As an example, Table 5.1 shows a list of FDA approved drugs for 2023 which show novelty in target or mechanisms of action (adapted from Avram et al. 2024).

When estimating the numbers of targets for approved drugs a tacit assumption is made that most approved drugs will affect a single target, which is certainly not the case in reality. Multiple target engagement may be problematical, leading to undesirable adverse drug reactions, but advantageous if giving better control of the disease. Another potential benefit is through drug repurposing to take advantage of another target in a different disease process. As with much of the work discussed in this book, machine learning will have an impact, in this case through mining large

Table 5.1 Novel drugs approved in 2023

Drug name	Type	Target	Indication
Fezolinetant	SM	Neurokinin 3 receptor	Vasomotor symptoms
Iptacopan	SM	Complement factor B	Haemoglobinurea
Nirogacestat	SM	Gamma-secretase	Desmoid tumours
Momelotinib	SM	Activin Receptor type I	Myelofibrosis
Capivasertib	SM	AKT 1–3	Breast cancer
Talquetalmab	Mab	GPRC5D	Multiple myeloma
Concizumab	Mab	TFPI	Haemophilia A/B
Tofersen	ASO	SOD1 mRNA	ALS
Nedosiran	siRNA	LDHA mRNA	Hyperoxaluria type 1

Adapted from Avram et al. (2024) with permission

SM small molecule, *Mab* monoclonal antibody, *ASO* antisense oligonucleotide, *siRNA* small interfering RNA

datasets to build on the target numbers mentioned above and to identify new opportunities for repurposing.

5.2.2.2 Traditional Target Discovery

Traditional in this context means the tried and tested process of identifying specific targets by studying biological mechanisms. The latter could include cell division, activation, differentiation, whole tissue effects such as muscle contraction, or many other phenomena associated with disease. In pre-internet days many ideas came from reading the academic literature in hard copy journals that would appear at regular intervals in the mail. The early literature on cytokines, for example, contained descriptions of cellular phenomena in the immune system that presented opportunities for drug development, starting with interleukin-1 (IL-1) for inflammation.[2] IL-2, 3,4,5 and 6 followed for growth of T cells, myeloid cells, immunoglobulin E production, eosinophil and B cell growth respectively. All of these mechanisms are involved in important disease areas and all have approved biological drugs, albeit marketed many years after the initial discoveries were made.

The development of the internet and online literature resources, conference proceedings in video as well as text formats naturally increased the number of target opportunities. It is impossible to say exactly how many of these opportunities exist, but at least some idea can be obtained by collating data online. The author conducted a basic pre-machine learning survey in 2016 using the HGNC catalogue of human protein-coding genes and non-coding RNAs as a point of reference (Zanders 2016). As of September 2024, there are 19,260 protein-coding genes, 9137 non-coding RNAs and 14,463 pseudogenes (HGNC 2024) which is not much greater than the figures from 2015. Existing or speculative opportunities for identifying drug targets were collated by searching the PubMed database using the search terms *therapeutic target* and *drug target*. The references could refer to a system, like *ferroptosis* for example or an individual molecule or set of molecules. Figure 5.1 summarises the numbers of references to illustrate the increase in literature on the subject from the beginning of the twenty-first century to 2024 with a slight dip in 2022–23 possibly related to the overall slowdown of research due to the COVID-19 pandemic.

Of the many examples of PubMed entries produced from the search term *therapeutic target* the following from Fu et al. (2023) shows how ideas for drug discovery programmes could be generated. The title and part of the Abstract should be self-explanatory:

Fibrinogen-like protein 2 in inflammatory diseases: A future therapeutic target The immunomodulatory role of FGL2 is evident in fibrin deposition-associated inflammatory diseases and cancer, suggesting that FGL2 expression could be exploited as a disease biomarker and a therapeutic target.

[2] There are currently at least 40 known interleukins (Dabbagh-Gorjani 2024)

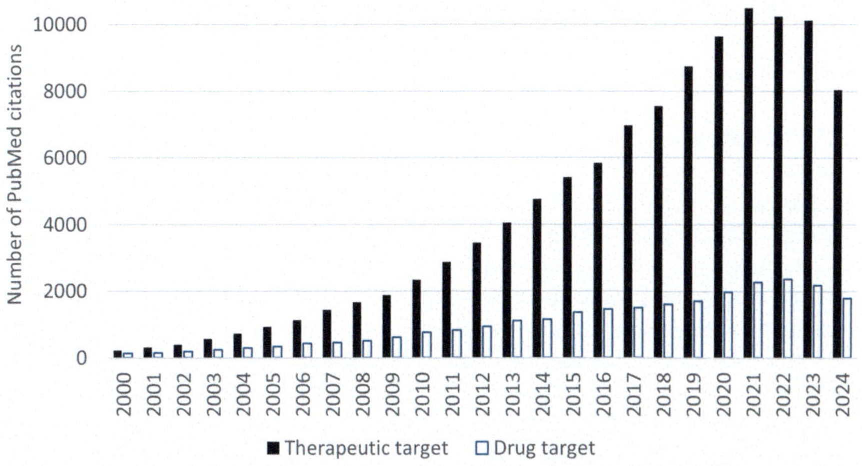

Fig. 5.1 Number of PubMed citations by year using search terms *therapeutic target* and *drug target*

A laborious manual curation of the literature entries found in the above searches was made in 2015/2016; papers related to individual protein and RNA targets resulted in 5329 entries relating to 9641 protein coding genes and 169 non-coding RNAs. While this is plenty of material to develop further, the process of generating the data is very inefficient and incomplete, in contrast to the power of machine learning, where data scraped from the literature can be analysed rapidly, at scale, and importantly used to reveal connections between different data elements.

5.2.2.3 The Role of Serendipity

A number of successful medicines were originally discovered accidentally by investigators who acted on Pasteur's famous words: *Dans les champs de l'observation le hasard ne favorise que les esprits préparés* (In the field of observation, chance only favours the prepared mind). Mention has already been made of the discovery of cis-platin drugs (see Chap. 2), other examples being penicillin and the repositioning of sildenafil for erectile disfunction by Pfizer. An overview of serendipity in drug discovery can be found in Ban (2022). Will serendipitous discoveries keep occurring, or will the processing of enormous datasets anticipate much of what could have been discovered by accident? Cis-platin and penicillin were discovered through laboratory experiments and not by using computers; unless a robot scientist can fully replace a human investigator, it seems reasonable to assume that accidental discoveries will still be made in the years ahead.

5.3 Developments in Cell and Molecular Biology

5.3.1 Genomics

5.3.1.1 Strategies

Have genetic strategies[3] been effective for increasing drug discovery success? The the approaches include *Mendelian Randomisation* and *Genome-Wide Association Studies (GWAS)*, both of which will be covered later. King et al. (2019) have shown that targets linked with coding variants are at least twice as likely to generate approved drugs after greater success in Phase II and III. In more recent work, drug mechanisms with genetic support are 2.6 times more likely to be successful (Minikel et al. 2024). An example of successful drug development based on genetic data is the proprotein convertase subtilisin/kexin 9 (PCSK9) gene associated with hypercholesterolaemia (Della Badia et al. 2016). Alirocumab and evolocumab are humanised monoclonal antibodies and inclisiran, a *small interfering RNA* (siRNA), all available as inhibitors of PCSK9 in situations where statin drugs fail to control cholesterol levels. Successful examples like these are still comparatively rare, highlighting the need for greater efficiency in genome-based target discovery.

Advances in laboratory technology and computer science underpin more ambitious target discovery programmes that until recently were essentially impossible to contemplate. Two logical approaches are possible: firstly, a detailed analysis of diseases of interest using genomic, proteomic and metabolomic data among others to identify actionable drug targets (diagnostics are covered briefly in Chap. 7). Secondly, a *human first, no hypothesis strategy* where multiple datasets are examined in an agnostic way to provide clues about disease associations and potential targets that may emerge as a result (discussed by Eisenstein, 2022a, b). The basic raw material, i.e. DNA sequences) required for either of these strategies is summarised graphically in Fig. 5.2 and covered in more detail in the following sections.

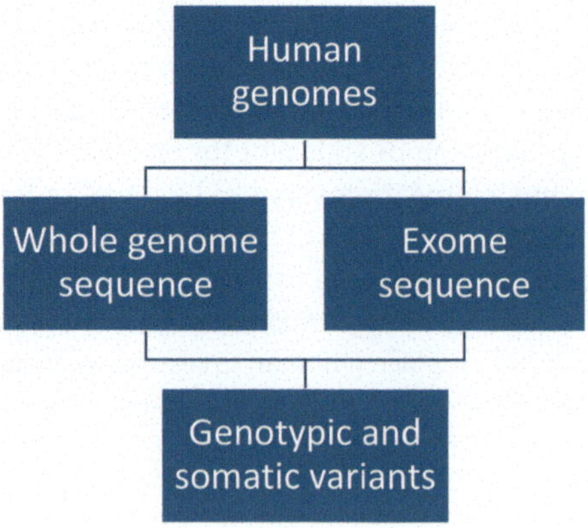

Fig. 5.2 Genomic source material for analysis

[3]The names genomics and genetics, while distinct subjects, are somewhat loosely used here

Human Genome Sequences

Although formally completed in 2003, The Human Genome Project covered about 92% of the roughly 3 gigabase sequence of human diploid DNA. The last 8% has now between sequenced by the Telomere to Telomere consortium (T2T) using technologies developed in the intervening years, so the sequence is now complete (Nurk et al. 2022). The male Y chromosome was sequenced later using another source of DNA since the CHM13 cell line used for the full sequence had an XX karyotype (Rhie et al. 2023). The sequencing technology will be described later, but a key point is the diversity of genomes used to establish sequence variations. Earlier studies were biassed towards samples from people of European descent who make up only about 15% of the global population. Lack of sequence data on the remaining 85% would mean missing novel DNA variants that could account for geographical variations in diseases and potential drug targets. However, this situation is changing rapidly, a recent example being the sequencing of 245,388 whole genomes from diverse US populations where 77% were from historically under-represented communities in biomedical research and 46% from racial and ethnic minorities (The All of Us Research Program Genomics Investigators 2024). WGS data from the UK Biobank include 150,000 sequences from predominantly British/Irish donors and a smaller proportion from Africans and South Asians, which reflects the population of the British Isles (Halldorsson et al. 2022). African genome sequences currently represent a small fraction of those available, so an initiative is underway to sequence 500,000 genomes from Africans and African Americans to help redress the balance (The Diaspora Human Genomics Institute 2025). Chinese sequencing efforts include a whole genome database of ethnic minority populations in Guizhou Province, Southwest China (He et al. 2024a) and Han Chinese genome resources. The Han Chinese are the largest ethnic group in East Asia comprising approximately 1.23 billion people approximating about 20% of the global population (Zhang et al. 2021). Other Asian sequencing projects referenced in Zhang et al. (2021) are Pan Asian, Japanese and Singaporean. Other sequencing efforts include Trans-Omics for Precision Medicine (TOPMed 2025) in the USA which has elements in common with the UK Biobank but focusses on heart, lung, blood, and sleep disorders (TOPMed 2025). In November 2023, the UK Biobank announced the release of 500,000 whole genome sequences, a significant achievement, taking 350,000 hours of genome sequencing, and costing over £200 million of investment from the UK Government and funding from Amgen, AstraZeneca, GSK and Johnson & Johnson (UK Biobank 2023). From the above, it can be seen that potentially millions of whole genome sequences will become available, perhaps even billions if the technology can be made more efficient and cost-effective. In the meantime, exome sequencing is a powerful way of identifying variants in the protein-coding regions that make up about 2% of the human genome. This can focus attention on potential drug target genes at scale, although important regulatory regions in the genome will of course be excluded. A recent example is the Regeneron Genetics Centre Million Exome dataset (RGC-ME) where 983,578 exome sequences have been analysed, albeit from predominantly European-origin donors (Sun et al. 2024).

Finally, the increasing volume of whole genome and exome sequence data are being made accessible to researchers worldwide through a variety of portals, one of which is the Genome Aggregation Database (gnomAD 2025).

The Dark Genome

The so-called *dark genome* represents about 98% of the human genome sequence but whose function is poorly understood, while the better characterised protein-coding regions comprise only about 2%. However, during evolution, retrotransposons have been incorporated into the genome using a *copy and paste mechanism* to insert RNA intermediates consisting of endogenous retroviruses no longer active in humans,[4] and LINE-1 (long interspersed nuclear element-1) (Baldwin et al. 2024). LINE-1 encodes two proteins ORF1p and ORF2p that effect the cut and paste mechanism, the latter with endonuclease and reverse transcriptase activities along with three domains with unknown functions. Because LINE-1 s can be reactivated and contribute to disease, ORF2p has been implicated in the pathophysiology of cancer, autoimmunity and ageing, making it a potential therapeutic target. The field has attracted the interest of companies such as Transposon Therapeutics, Rome Therapeutics, Evaxion Biotech and Enara Bio (Eisenstein 2024).

In addition to retrotransposons, the dark genome contains a large number of non-coding RNAs, both small (<200 nucleotides and long (>200 to thousands). Drug discovery efforts have mainly focused on the small RNAs, but there are more than 15,000 *long noncoding RNAs* (lncRNAs) some of which are involved in cancer, inflammation and other diseases. They act as signals, decoys, guides and/or scaffolds to modulate transcription, translation, splicing patterns, protein/RNA stability and chromatin remodelling (Schiesser et al. 2022). Drugs targeting these noncoding RNAs will be covered in Chap. 6.

5.3.1.2 Sequencing Technology

Large scale DNA sequencing projects require technologies that operate at high throughput with accurate reads. Since one of the objectives is to identify variants that may differ by a single nucleotide, this accuracy is of paramount importance.

First Generation Sequencing

The archetypal first DNA generation sequencing method was devised by Frederick Sanger who used radioactive nucleotides to image DNA fragments separated on polyacrylamide gels. The method was effective but time consuming, so it was later modified by Applied Biosystems to use fluorescent nucleotides as labels and capillary gel electrophoresis to separate the fragments and register them in a computer. The system was used to create the first human genome sequences in 2003.

[4] Unless activated in cancer leading to neoantigens that may be recognised by the immune system (see Eisenstein 2024)

Second Generation Sequencing
Second generation (or Next Generation) sequencing began with the introduction of pyrosequencing by Roche/454 Life Sciences in the early 2000s. Pyrosequencing involves the copying of an immobilised DNA template by DNA polymerase and release of pyrophosphate when the correct nucleotide (A,C,G or T) is incorporated. The pyrophosphate release is detected using a luminescent assay. Alternatively, *sequencing by synthesis* used massively parallel arrays of single DNA strands immobilised on a flow cell and used as templates for DNA polymerase to allow incorporation of the four nucleotides labelled with a different fluorochrome. The originator company Solexa was acquired by Illumina in 2007 which now markets sequencers using this technology.

Third Generation Sequencing and Beyond
New technologies are capable of reading long sequences without compromising on accuracy. The full human genome sequence described by Nurk et al. (2022) was generated as follows (coverage in brackets): PacBio's HiFi (30×), Oxford Nanopore Ultralong-read sequencing (120×), Illumina PCR-free sequencing (100×) and Arima Genomics Hi-C (70×).

The PacBio HiFi system uses Single Molecule, Real-Time (SMRT) sequencing in the form of a microchip with an array of immobilised DNA samples. Incorporation of fluorescent nucleotides during primer extension is measured by light emission (PacBio 2025). The Oxford Nanopore technology uses a nanoscale protein pore embedded in an electrically resistant polymer membrane (Oxford Nanopore 2025). A constant voltage is applied to drive negatively charged single-stranded DNA or RNA molecules. The speed is controlled by a motor protein that ratchets the nucleic acid molecule through the nanopore in a stepwise manner. Changes in the ionic current during translocation correspond to the nucleotide sequence present in the sensing region (Wang et al. 2021a). Illumina's PCR-free sequencing is used to avoid the use of PCR which can cause errors when amplifying genomic DNA immobilised on beads (Illumina 2025). Arima Genomics Hi-C captures the organizational structure of chromatin in three dimensions, where genomic sequences that are distal to each other in linear distance can be closer to each other in the 3-D space. It analyses spatial genome organization and maps higher-order chromosome folding and topological associated domains (Arima Genomics 2025).

Issues for Future Sequencing
Large scale genome sequencing projects have been made possible by the technical developments outlined in the previous sections. Long read technology coupled with increased accuracy (aided by machine learning) is transforming the field; cost is becoming much less of an issue, as the price for sequencing a full human genome is now below $1000. Competition between the manufacturers of instruments and sequencing reagents is driving the price down to the extent where Ultima Genomics (Ultima 2025) is proposing a $100 sequence, while others such as Illumina are making the process affordable for all. Healthcare providers such as Britain's National Health Service will initially offer WGS to individuals with rare diseases

or cancer (NHS Genomic Medicine Service 2025). With similar rollouts across the globe, the number of WGS reads should increase massively, thereby generating exabytes of data. The real issues for the future, therefore, will not be the sequencing *per se*, but having sufficient compute architecture plus the bioinformatics required to make sense of the data. Sequence is just sequence after all, what it encodes and the variation between individuals is what drives personalised medicine, target discovery etc.

5.3.1.3 Sequence Variants

Most phenotypic variation between individuals is due to DNA sequence changes leading to modified gene expression and splicing (Taylor et al. 2024). These mutations may reside in the germline mutations of the human genome, or in the case of genetic diseases such as cancer, be somatic within specific tissues. *Mutational signatures*, obtained by using DNA sequencing or targeted arrays, represent the variations between individuals with and without a disease of interest and are listed in Table 5.2.

The signatures listed in Table 5.2 are described in more detail by the Sanger Centre (COSMIC Mutational Signatures 2025) with a focus on cancer signatures created through somatic mutations in tumour cells with SNPs and Microsatellites being germline mutations useful for disease gene mapping (Li et al. 2004). The numbers of variants identified by comparison with the human reference genome[5] have been presented in 2015 (The 1000 Genomes Project Consortium 2015). DNA from 2504 individuals from Africa East Asia, Europe, South Asia and the Americas was analysed by low-coverage whole-genome sequencing, deep exome sequencing, and high-density SNP microarrays. A typical genome differs from the reference human at genome at 4.1 million to 5.0 million sites, 99.9% being SNPs and short indels. There were an estimated 2100 to 2500 structural variants affecting 20 million

Table 5.2 List of mutational signatures in human genome

Name	Variant type	Comment
SBS or SNV	Single base substitution or single nucleotide variant somatic mutations	
DBS	Double base substitution	
ID	Insertion/deletion (Indels)	Insertion/loss of 1–50 NTs
CN	Copy number variations	
SV	Structural variations	Large variations >1Kb
RNA SBS	RNA single base substitution	
SNP	Single nucleotide polymorphism	Germline mutation in >1% population
Microsatellites	1-6 bp repeats	No. repeats vary between individuals

bp base pair, *NT* nucleotide

[5] Preceding the one published by Nurk et al. (2022) which contains about 200 Mb more sequence for analysis

bases of sequence (1000 large deletions,160 copy-number variants, the remaining being microsatellite insertions (Alu, LINE1) and a small number of inversions). Most variants in a single genome are common, with 1–4% having a frequency of 0.5%. The authors suggest that deep sequencing the entire sample would at least double the total number of variants in the sample but increase the number of variants in a typical genome by only 20,000–60,000. This brings up the subject of deep mining the human population for the common and rare variants required for fine mapping of disease genes and drug target selection. Whole-genome sequencing is ideal and will become routine for analysing large populations, but it is still relatively expensive and technically challenging for large scale analyses. As a compromise, genotypes for unmeasured variants are generally inferred using a procedure known as *genotype imputation* allowing exploration of previously unattainable sample sizes (Naito and Okada 2024).

5.3.1.4 Mapping Disease Genes

Mendelian Diseases

Mendelian conditions are caused by a single gene that is passed through the generations as first described by Gregor Mendel in the nineteenth century. Thousands of monogenic diseases exist including those with actionable drug targets such as cystic fibrosis and sickle cell anaemia. The Online Mendelian Inheritance in Man (OMIM 2025) database, as of second October 2024, lists 6501 phenotypes of single gene disorders connected with 4570 genes plus 236 somatic mutation phenotypes associated with 129 genes (OMIM 2025). The growth of this database has been made possible through advances in exome sequencing that has led to a dramatic increase in gene discovery in the last decade (see Bamshad et al. 2019).

While the investigation of Mendelian traits has proved successful for many monogenic diseases, the large amount of genetic variation associated with complex diseases like asthma and neurological conditions is more problematic (Claussnitzer et al. 2020). Exploiting this genetic variation for disease mapping and drug target identification is covered in the following section.

Genome-Wide Association Studies (GWAS)

Genome-wide Association Studies (GWAS) identify large numbers of genetic variants across the genome and determine their association with a phenotype of interest. Early studies revealed common variants with more subtle effects than with a monogenic disease, but now improvements in sequencing coverage and analysis have extended GWAS to rare variants as well. It is clear that common diseases of interest to the biopharma industry are not driven by a few Mendelian traits, but by thousands of variants with individually modest risk. A good illustration of this is provided by the GWAS of human height, a highly hereditable trait that is readily measurable. Yengo et al. (2022) identified 12,111 SNPs associated with height in a sample of 5.4 million people of diverse ancestries and suggest that this is the upper limit of what could be found, at least in Europeans.

Obviously complex late-onset diseases are likely to be more complex in terms of genomic variants, to quote Yengo et al. (2022): *if these predictions reflect proportional levels of saturation between traits, then we could expect that two- to tenfold larger samples would be required for GWASs of inflammatory bowel disease (×2, that is, n = 10 million), schizophrenia (×7; n = 35 million) or BMI (×10; n = 50 million) to reach a similar saturation of 80–90% of SNP-based heritability.*

Apart from the challenge of obtaining enough samples, actually connecting a variant to a functional gene defect (and therefore potential drug target) may not be straightforward. GWAS variants often occur in highly correlated groups[6] near protein-coding genes and act in a regulatory capacity, affecting either transcript or protein concentration rather than protein function itself. Although most GWAS identify protein-coding genes, there can be no *a priori* assumption that this is the case for a causal gene driving a GWAS signal (Hukerikar et al. 2024).

Quantitative Trait Loci (QTL)

A *quantitative trait locus* (QTL) is a region of DNA associated with a specific phenotype or trait that varies within a population. The variance is continuous, like height as opposed to discrete, such as eye colour. QTL mapping is a statistical analysis to identify which molecular markers lead to a quantitative change of a particular trait. Since a single locus may include many variants, imputation or whole-genome sequencing is a key prerequisite for QTL mapping to enable precise identification of the contributing molecular marker. The different types of QTL listed in Table 5.3 relate to the nature of the phenotypic change (see Illumina Techniques for example 2025).

eQTLs are *expression QTLs* that act on locally (cis) or at a distance, possibly on a different chromosome (trans). meQTL, caQTL and bQTL affect gene expression at the level of transcription for example, through chromatin accessibility and transcription factor binding, while pQTLs affect the levels of protein expression.

Mendelian Randomization (MR)

GWAS of complex diseases can produce many thousands of variants in mostly non-coding regions of the genome, each contributing a small amount of risk. Variants will be either directly or indirectly associated with a drug target of interest

Table 5.3 Different types of QTL

cis-eQTL	Expression QTL	Local action
Trans-eQTL	Expression QTL	Distant action
meQTL	Methylation QTL	Methylation patterns
caQTL	Chromatin accessibility QTL	Nucleosome packing
bQTL	Binding QTL	Transcription factor binding
pQTL	Protein QTL	Protein expression levels

[6]i.e. in *linkage disequilibrium*

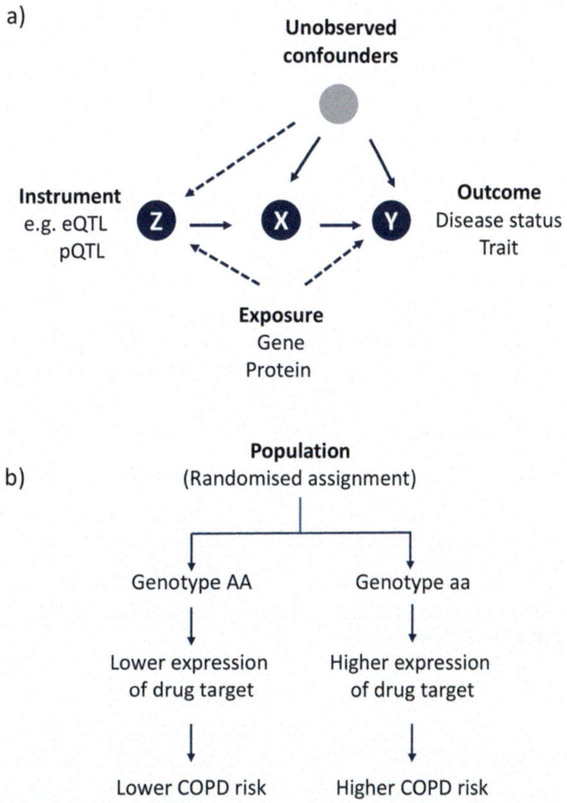

Fig. 5.3 (**a**) Instrumental variables. The instrumental variable (Z) can be based on DNA sequence or protein expression and directly influence the disease outcome or trait (Y). The instrument is unaffected by unobserved confounders or exposure to (possibly environmental) gene and protein actions (illustrated a dotted lines). (**b**) Mendelian association of genes affecting COPD risk within a population. (Wang et al. 2024 under CC-BY-4.0)

thus raising the need to distinguish between correlation and causation.[7] In *Mendelian randomization (MR)*, genetic variants produced at conception and randomly assigned to the resulting offspring are analysed statistically (Schmidt et al. 2020). The genetic variables, which may be DNA based or protein QTLs, are used as *instrumental variables (IVs)*, or just *instruments* (Michoel and Zhang 2023).

The work of Wang et al. (2024) on chronic obstructive pulmonary disease (COPD) provides an example of how the above genetic studies work in practice and includes references to modern databases containing key information on eQTLs, proteins and drug targets. COPD is a major cause of morbidity and mortality, with existing medicines providing only symptomatic relief. GWAS have been used to identify SNPs associated with the disease but without providing clear drug targets, so MR was employed on eQTL and p-QTL for 4317 druggable genes. Their expression in whole blood (exposure) should modify the risk of developing COPD (outcome). According to Mendel's law, a population randomly assigned a genotype with high expression of a particular gene will be exposed to high expression of a particular gene for life. These points are illustrated in Fig. 5.3.

[7] Amusing examples can be found in Spurious Correlations (2025), e.g. a strong correlation between distance of Jupiter to Sun and number of secretaries in Alaska

Table 5.4 Databases used by Wang et al. (2024)

Name	Tissue/diagnosis	Reference(s)
eQTLGen consortium	Whole blood	Võsa et al. (2021), eQTLGen consortium (2025)
GTEx consortium	Lung tissue	Lonsdale et al. (2013), GTEx Project (2025)
deCode genetics/ Amgen Inc	Plasma proteome	Ferkingstad et al. (2021)
Finngen consortium	Diagnosed COPD	Finngen (2025)
SpiroMeta consortium	FEV levels	Obeidat et al. (2021)
UK biobank	FEV levels	UK Biobank (2023)
Druggable genome	Survey of drug targets and variants	Finan et al. (2017)
AZ Phenomics portal	UK Biobank phenotypes and exomes	Wang et al. (2021b)

The Wang et al. (2024) study used data from the databases or studies listed in Table 5.4 and mentioned in the following sections.

Exposure Data

eQTLs were obtained from 25,482 whole blood samples (eQTLGen Consortium 2025) and postmortem lung tissue from 515 donors (GTEx consortium). From these samples all conditionally independent eQTLs in whole blood and lung tissue, which covered 2531 of the 4317 druggable genes, could be identified in at least one tissue. pQTLs were obtained from whole blood for 4907 plasma proteins (Ferkingstad et al. 2021).

Outcome Data

COPD data were presented as doctor-diagnosed disease or lung spirometry (FEV1) measurements taken during random screening. The former were taken from the Finngen consortium and consisted of 16,410 patients and 283,589 controls.

FEV 1data were taken from the UK Biobank as well as the SpiroMeta consortium.

Summary Data Mendelian Randomisation (SMR)

SMR estimates the effect of exposure (genes targetable for drug intervention at the transcriptional or protein expression level) on the outcome (e.g. COPD diagnosis) using genetic variation significantly associated with the exposure. More detailed information on how the analyses were performed is given in the Wang (2024) paper, including protein SMR, leading to a list of druggable genes significantly associated with diagnosed COPD or compromised FEV1. The number of genes listed in Table 5.5, gives an idea of the challenges involved in moving from identifying a list of candidates to choosing interventions for a clinical development programme. Some of the gene products are targets for existing medicines (in brackets), namely

Table 5.5 Druggable genes significantly associated with COPD

Significant association with diagnosed COPD		
C2		
GPC2		
Significant association with spirometry defined FEV1		
TESK2	ERBB3	MMP15
AKR1A1	PSMA4	MRC2
LMCD1	MAST2	BMP4
SERPING1	CHI3L2	LTK
TYRO3		
Significant association with FEV1		
APH1A	TESK2	LRPAP1
CHI3L2	ADAM33	LTBR
COL7A1	COL15A1	NPNT
GDF11	LINGO1	SULT1A1
GDF5	PTK7	RASGRP3
HSP90AA1		

From: Wang et al. (2024) under CC-BY-4.0

MMP15 (Montelukast), PSMA4 (Maroizomib), APH1A, (Begacestat etc) and ERBB3 (Duligotuzumab etc) so there is potential for drug repurposing. Other targets will have a role in the pathophysiology of COPD and must be evaluated in a similar way to drug discovery programmes for other conditions.

Cancer Genomes

The above sections on DNA variants have centred mostly on hereditable germline mutations that lead to monogenic or polygenic diseases. Cancer, however, is a highly unstable genetic disease resulting in the main from somatic mutations within specific tissues. This makes the situation more complicated since individual patients will have specific mutations that influence the aggressiveness of the tumour and its susceptibility to therapeutic interventions. This of course is the realm of personalised medicine where technical improvements in high throughput genome analysis and machine learning algorithms will contribute to future advances in cancer therapy and diagnostics.

Mutational signatures from a cancer genome survey encompassing most cancer types revealed 49 single-base-substitutions (SBS), 11 doublet-base-substitution (DBS), 4 clustered-base-substitution and 17 small insertion-and-deletion signatures (indels) (Alexandrov et al. 2020). The field of cancer genomics is understandably vast, but the following cases illustrate recent findings relevant to drug discovery and clinical medicine.

Mutagens and Geographical Location

To understand how mutagens and geographical locations influence mutation patterns in cancers, it is necessary to conduct large scale WGS to reveal patterns of single base substitutions (SBS). The Cancer Grand Challenges Programme includes a

Mutagraphs team, whose aim is to put these ideas into practice (Cancer Grand Challenges Mutagraphs 2024) and who published a study of 962 clear cell renal cell carcinomas from 11 countries with varying incidence (Senkin et al. 2024). The somatic mutation profiles differed between countries and included SBS variants involved with deamination of 5-methylcytosine and cytosine deamination by apolipoprotein B mRNA-editing enzyme, catalytic polypeptide-like (APOBEC). In Romania, Serbia and Thailand, mutational signatures characteristic of aristolochic acid compounds were present in most cases, but these were rare elsewhere and In Japan, a mutational signature of unknown cause was found in more than 70% of cases but in less than 2% elsewhere. Known signatures of tobacco smoking correlated with tobacco consumption, but not with obesity or hypertension, The results of this study reveal the existence of multiple geographically variable mutagenic exposures and highlight the value of large scale WGS in cancer research.

Mutational signatures are clearly important in understanding the action of different environmental stimuli on normal cells, but paradoxically, cells can maintain a high mutational burden without becoming cancerous. It is possible that tumour promoters (chemicals radiation etc) *tip the cell over the edge* leading to uncontrolled proliferation and carcinogenesis. This important area for current and future study is the remit of Cancer Grand Challenges-Normal Phenotypes (2025).

Colorectal Cancer

Colorectal carcinoma (CRC) is of major clinical significance, being the third most prevalent cancer after that of breast and lung. WGS of 2023 CRC samples from the UK 100,000 Genomes Project has produced a detailed landscape of somatic mutations providing a detailed insight into CRC pathology (Cornish et al. 2024). More than 250 putative CRC driver genes were identified, many not previously implicated in CRC or other cancers, and bearing several mutations in non-coding regions. CRC could be divided into subgroups with independent prognoses:

MSI (microsatellite instability-positive, mismatch repair deficient)
POL (DNA polymerase ε proofreading-deficient)
MSS (microsatellite-stable)
SBS93 single base substitution (smoking related).

Escape mutants, often in the form of HLA copy number changes, are near ubiquitous in hypermutant tumours adding to the problem of immune evasion in cancer. Many driver mutations are potential drug targets however, so there are plenty of new therapeutic opportunities.

Metastasis

Metastasis is responsible for the vast majority of cancer-related deaths and is therefore a high priority target for therapeutic intervention. Studies using WGS to compare primary tumour mutation profiles with metastasised cells, in principle should identify actionable drug targets.

Studies of this type have been published, for example by Martínez-Jiménez (2023) who show some genetic differences between the two tumour stages, but these include the results of chemotherapy on DNA integrity. It would appear that with both genetic and non-genetic mechanisms driving metastasis, the situation is more complex (Karras et al. 2024). From their review of the literature it would appear that metastasis is rarely initiated by a genomic alteration but through intensification of oncogenic driver genes such as CDKN2A, MYC, TP53 and those involved in therapy resistance. Interestingly, metastasis only rarely occurs spontaneously in mouse models of cancer as they often do not have the genetic complexity of the human disease under investigation. Metastasis often involves changes in cell plasticity and epigenetic control of gene expression, a topic to be covered in the next section.

5.3.1.5 Functional Genomics- Gene Expression

DNA may be a target for certain chemotherapeutic drugs, but it is of course the template for transcription into mRNA (and other RNAs) and translation into the proteins that constitute the majority of drug targets. Transcription is highly regulated through the interplay of DNA elements, chromatin structure and RNA modifications. This is summarised in Fig. 5.4 which focuses on the *transcriptome* or genome-wide complement of transcribed RNA.

Fig. 5.4 Gene expression

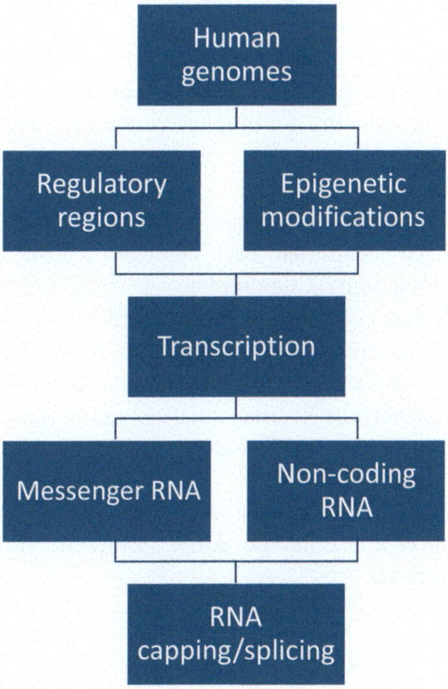

Transcriptomics

Transcriptomics involves the cataloguing of all types of DNA transcript, mRNAs, long non-coding RNAs (lncRNAs) and micro RNAs (miRNAs) and to identify splicing and modifications. RNA expression under different conditions, including those caused by drug treatment, is measured using methods developed over many years, such as northern blotting and reverse transcription PCR (RT-PCR), but these are low-throughput approaches.

To address this problem, microarrays were developed for detecting thousands of RNA transcripts in parallel using glass slides or photolithography to immobilise DNA probes. Microarrays have been, and still are, invaluable for basic biological investigations and drug discovery but are being supplemented by *RNA-Seq* using Next Generation Sequencing (NGS). This is performed by fragmenting cDNA copies of bulk RNA then ligating with adapters to produce NGS sequencing libraries (detailed in Wang et al. 2009). The sequence of each fragment can then be aligned to a reference genome (or transcripts) or assembled into a transcript map that shows the structure (e.g. splicing) and level of expression. Since the method is based on sequence, the transcriptome can be surveyed at single base resolution and a genome-level scale.

However, technology moves on, from analysing mRNA expression in bulk cells and tissues to that in single cells using single cell RNA sequencing (*scRNA-seq*). According to Aldridge and Teichmann (2020), the technique was first developed in 2009 using a small number of mouse cells. Since then, scRNA-seq has developed into a significant tool for examining transcription at an unprecedented level of resolution. Given the importance of this technique and likely future developments, the basic procedures for scRNA-seq are shown in Fig. 5.5 and following:

The figure shows a basic procedure that has been improved by using microfluidics with nanolitre oil droplets containing single cells (summarised by Macosko et al. 2015):

1. prepare a single-cell suspension from a tissue
2. co-encapsulate each cell with a distinctly barcoded microparticle (bead) in a nanolitre-scale droplet
3. lyse cells after they have been isolated in droplets
4. capture a cell's mRNA on its companion microparticle, forming STAMPs (single-cell transcriptomes attached to microparticles)
5. reverse-transcribe, amplify, and sequence thousands of STAMPs in one reaction
6. use the STAMP barcodes to infer each transcript's cell of origin.

Clark et al. (2023) have developed a microfluidics-free workflow to improve throughput, and no doubt further developments will be described in the years to come.

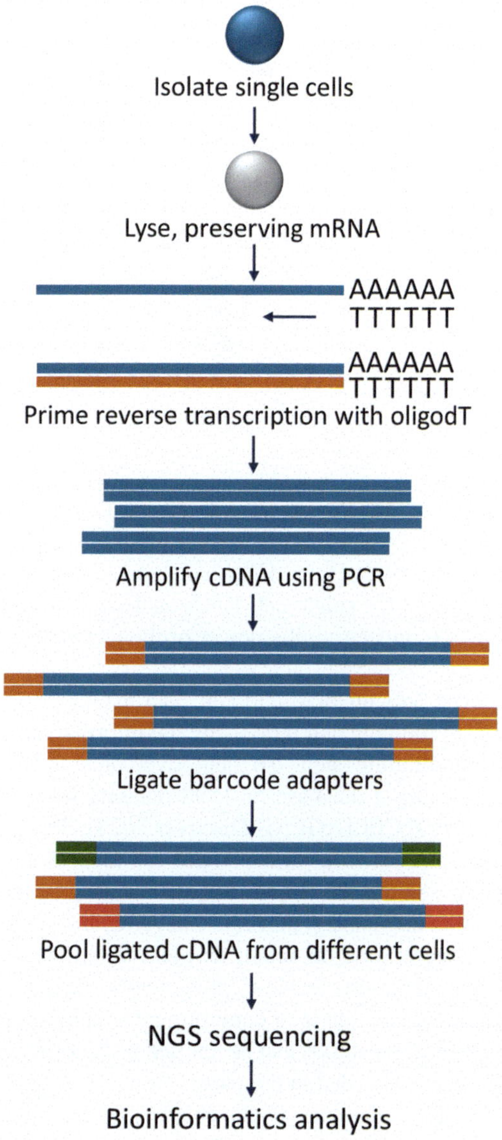

Fig. 5.5 Basic outline of scRNA-seq. Single cells (obtained by cell sorting or otherwise) are lysed to preserve mRNA which is reverse transcribed using oligodT as a primer. The resulting cDNA is amplified using PCR then ligated with adapters containing a unique barcode to identify the originating cell. The adapted libraries from different cell samples are pooled and subjected to NGS to identify the cell of origin and the mRNA using a bioinformatics workflow

Sequence Data Processing

The NGS sequence data emerging from a scRNA-seq experiment must be processed in a bioinformatics pipeline. Illumina and other genomics companies provide the relevant software to clean the data, identify transcripts and assign them to a cell of origin. The resulting high-dimensionality data must be reduced and displayed in a form that enables visualisation in two or three dimensions. The most common statistical techniques used are *t-distributed Stochastic Neighbour Embedding*

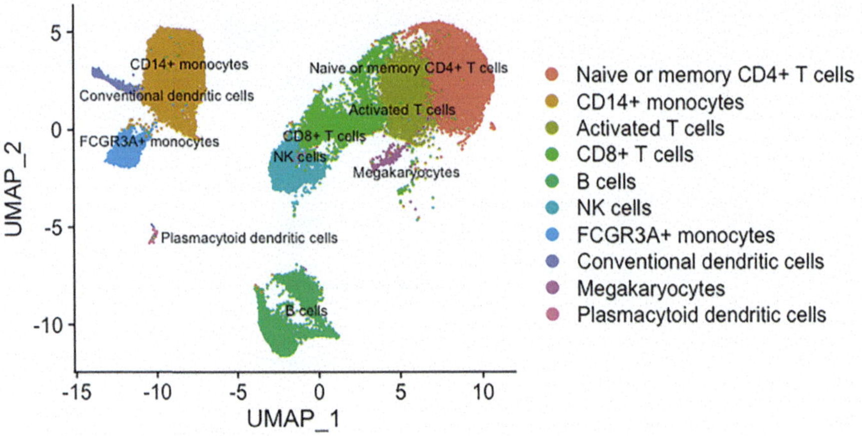

Fig. 5.6 scRNA-seq of leukocyte subsets showing results of UMAP projection. (Picture courtesy of Harvard Chan Bioinformatics Core under CC-BY-4.0 2025)

(t-SNE) and *Uniform Manifold Approximation and Projection (UMAP).* A UMAP projection for single cell RNA analysis of different leukocyte subsets is shown in Fig. 5.6 illustrating the power of the technique.

The amount of single cell data available worldwide is increasing to the point where sourcing and processing the information for various applications is becoming a rate-limiting step. To help overcome this, the Chan Zuckerberg Initiative is supporting a database of 1598 datasets from 916 cell types, totalling 96 million cells (CZ CELLxGENE Discover 2025). The database allows users to analyse cell and gene networks at the single cell level and generate ideas that can among other things, be translated into drug discovery programmes.

Van de Sande et al. (2023) have reviewed the application of scRNA-seq to drug discovery in some detail; as most of the authors are from major pharmaceutical companies, this review presumably reflects some of the thinking in these organisations and their commitment to scRNA-seq and related technologies. Details of how the methods can be applied to the different stages of discovery and development are provided in the paper.

For the purposes of this chapter, an example of the practical implementation of scRNA-seq In conjunction with deep learning is provided by Theodoris et al. (2023) who developed Geneformer, a deep learning model pretrained on about 30 million single-cell transcriptomes.[8] This enables the mapping of networks whose core regulatory elements are likely to prove better drug targets than peripheral effectors acting downstream. Geneformer is based on *transfer learning* and a *self-attention mechanism*, the former where generalised data (transcriptomic here) can be applied

[8] The CZ CELLxGENE Discover database was not used in this study

to many different situations; self-attention mechanisms can examine large input spaces and learn which elements are most important to focus on in each context. As a result, Geneformer can enable predictions in settings with limited data.

Details of the methodology and results can be found in Theodoris et al. (2023). Briefly, gene expression in each cell is normalised to a rank value which passes through six transformer encoder units composed of a self-attention layer and feed forward neural network layer. During pretraining, 15% of the genes within each transcriptome were masked, and the model was trained to predict which gene should be within each masked position in that specific cell state using the context of the remaining unmasked genes. This has the advantage that it is entirely self-supervised and can be accomplished using completely unlabelled data. The Geneformer model embeds each gene into a 256-dimensional space encoding the characteristics of the gene specific to the context of that cell.

As an example of its use, the model could mimic *in silico* the reprogramming of fibroblasts into *induced pluripotent stem cells (iPSCs)* by the transcription factors OCT4, SOX2, KLF4 and MYC. By adding these four genes to the front of the ranked encodings for fibroblasts it could shift the virtual transcriptome towards an iPSC state. Other findings included predictions of gene dosage effects (important in interpreting GWAS and similar data), chromatin dynamics (see later) network dynamics, *in silico* gene network analysis and *in silico* treatment analysis. Lastly, to give an idea of what might be achieved by combining transcriptomics with deep learning algorithms for target (and off target) prediction, Theodoris et al. (2023) could identify target candidates for cardiac myopathy by applying *in silico* analyses of gene expression to diseased and normal heart tissue.

Transcriptional Regulation

The regulation of gene expression is central to developmental biology, e.g. through activation of embryonic genes and their repression during adulthood. In pathology, aberrant gene expression is a hallmark of diseases such as cancer and inflammation so understanding how gene expression is controlled is essential for both basic science and for developing drug discovery strategies. Understanding transcriptional regulation requires a knowledge of the *3-D genome* i.e. the regulatory sequences DNA (promoters, enhancers and repressors) plus chromatin structure and epigenetic modification of both DNA and histone proteins. Details of the different domains used to package DNA into the cell nucleus along with a discussion of the functions of the 3-D genome are given in Krumm and Duan (2019).

Figure 5.7 shows the basic organisation of chromosome looping when transcription factors are brought into proximity with components of the *pre-initiation complex* to initiate transcription. The latter is a complex group of proteins that positions RNA polymerase on the promoter region to allow transcription to proceed from one of the DNA strands. It requires DNA polymerase II plus the general transcription factors TFIIA, TFIIB, TFIID, TFIIE, TFIIF, and TFIIH. Other components for activation include the 1.2 MDa multiprotein complex *Mediator*, plus cofactors and gene-specific activators and repressors (Greber and Nogales 2019).

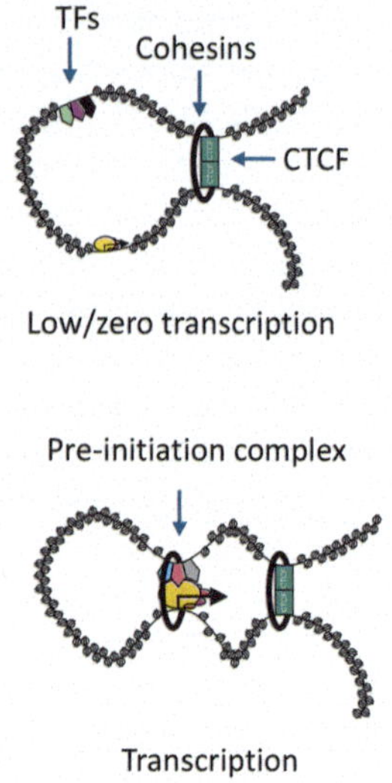

Fig. 5.7 Chromosome looping. Looping of DNA by cohesion complex and CTCF with low/zero transcription until further looping to bring transcription factors (TFs) into contact with the pre-initiation complex allowing transcription. (Adapted from Palstra and Grosveld 2012 under CC-BY-4.0)

Table 5.6 Some methods for chromatin-DNA/RNA analysis

Primary structure	Technologies
Sequence features	WGS, exome
DNA methylation	Bisulfite sequencing
DNA bound factors	ATAC-seq; ChIP-seq
Nucleosome positioning	SONO-seq; FAIRE-seq
Chromatin accessibility	DNAse-seq; ATAC-seq
Secondary structure	
Nucleosome-nucleosome interactions	No methods listed
Teritary structure	
Promoter-enhancer looping	Hi-C
Megabase-scale chromosome domains	Capture-C

From: Riska and Greenleaf (2015)

A genome wide analysis of regulatory protein-DNA interactions (short range, *cis* and long-range *trans*) helps in the interpretation of GWAS variants, most of which are encoded by non-coding regions in the genome. Different techniques have been developed for whole genome-level analysis of areas of active transcription, binding sites for transcription factors and polymerases and delineation of active enhancer regions. These are listed in Table 5.6, while a comprehensive review of the main techniques has been provided by Riska and Greenleaf (2015).

Variations and innovations in the area of 3-D genomics appear regularly in the literature and cannot be covered here in any detail. There is of course commercial activity in this area, particularly with a view to identifying novel drug targets from GWAS data. As an example, Enhanc3D Genomics Ltd. in the UK are using their GenLink3D platform, based on Hi-C, to visualise the folding of all genes and their enhancers simultaneously. This is designed to link enhancers and non-coding genetic variants to their target genes (Enhanc3D Genomics 2025), see also Javierre et al. (2016).

Much of the technology mentioned above was developed using bulk cells and tissues, but as with scRNA-seq, single cell methods can provide much finer detail and answer questions that cannot be addressed by methods that use unfractionated tissues. This is illustrated by Sundaram et al. (2024) who have examined single cell chromatin accessibility across eight tumour types derived from the Cancer Genome Atlas (TCGA 2025).[9] ATAC sequencing[10] was used to measure chromatin accessibility in individual cells from normal and diseased tissues, the idea being to define tumour cell-specific regulatory programmes without contamination from stromal and immune cells. The results showed that that the chromatin signature of a basal-like subtype of breast cancer is most similar to secretory-type luminal epithelial cells. Neural network models were then trained to learn a cancer-specific gene regulatory *grammar*. SNVs and mutations found in cancer could be introduced to the neural network and shown to be enriched near cancer-associated genes showing importantly, that somatic noncoding mutations can drive changes in cancer-associated genes.

The above example is not the only one where single cell chromatin analyses can provide useful information. Transcriptional enhancer activity in single cells can be detected by measuring the expression of bidirectional enhancer RNAs (eRNAs); these are relatively small (~500NTs) members of the long non-coding RNA family that is increasingly seen as being an important regulator of cell function, including gene expression (Batista and Chang 2013). By mapping the transcription start sites of mRNAs, long noncoding RNAs, and eRNAs in single CD4+ve T-cells, Oguchi et al. (2024) could measure gene transcription and enhancer activity simultaneously and identify rare subsets of CD4+ve T cells such as GPR25-expressing Treg cells. The fine level of detail obtainable using this and related techniques should prove useful for diagnostics and target discovery in general.

Lastly, the technique of *massively parallel reporter assays (MPRAs)* has been used to identify cis-acting DNA elements at high throughput using libraries engineered to drive a reporter gene carrying a unique DNA barcode in its 3' UTR. As with the other gene expression approaches, the assay system has been adapted for single cell analysis (*scMPRA*) as described by Zhao et al. (2023). They use a two-level barcoding system allowing single cell measurement of the copy number of all reporter genes present from mRNA levels alone. One barcode marks the cis-acting DNA sequence, and the other is a random barcode to assess the copy number of reporter genes in a single cell. Essentially, the number of different barcode pairs in a single cell becomes an effective proxy for the copy number of the cis-acting DNA elements in that cell. As a result, core promoter activity was

[9] The Cancer Genome Atlas Program (TCGA)

[10] Single cell ATAC-seq (scATAC-seq) using isolated cell nuclei and microfluidics in a similar way to scRNA-seq

broadly similar between K562 and HEK293 cells, but there were differences (11 upregulated and 59 downregulated) in K562 cells showing that the method could discriminate between cis-acting elements in different cell types. scMPRA could also be applied to different cells in the developing mouse retina showing its value in analysing complex tissues for differential cis-acting DNA elements.

The techniques described above for understanding transcriptional regulation at the level of non-coding DNA should help to decode the effects of the many sequence variants that map to diseases of interest. They may be drug targets in their own right but if not, at least they can be used to map the various regulatory elements in a network where key druggable effector targets may be uncovered.

Pre-mRNA Splicing

Previous sections have discussed DNA variants involved with transcriptional control of mRNA abundance through cis-acting DNA elements; these relate to the eQTLs used for relating GWAS data to specific diseases. However, there is another level of genetic regulation, pre-mRNA splicing (alternative splicing) which contributes to the expression of sQTLs, a separate class to eQTLs. The relative contributions of eQTLs and sQTLs to disease has been debated but recent studies appear to bear out the importance of the latter (see Qi et al. 2022) and Wang et al. 2023). Alternative splicing affects more than 95% of human genes to create multiple mRNA isoforms from a single pre-mRNA resulting in the translation of multiple protein isoforms. Aberrant splicing can contribute to a number of diseases including cancer and neurodegenerative disorders (Liu et al. 2022) and is therefore of biopharmaceutical interest.

As an example, Wang et al. (2023) determined splicing patterns in non-small-cell lung cancer (NSCLC) using RNA-seq data mapped to a reference genome to identify intron-exon boundaries.[11] Alternative splicing in 19 genes was significantly associated with NSCLC risk, one of which reduced the risk of disease by promoting FARP1[12] exon 20 skipping, thereby downregulating expression of the long transcript FARP1–011. Transfection of the long and short FARP1 transcripts into A549 and PC9 cells showed that the long transcript promoted migration and proliferation, thus functionally linking the sQTL to a disease mechanism.

Epigenetics and Genomics

To quote from Dupont et al. (2009):

Conrad Waddington introduced the term *epigenetics* in the early 1940s defining it as the branch of biology which studies the causal interactions between genes and their products which bring the phenotype into being. Today, the definition is generally accepted to be the study of changes in gene function that are mitotically and/or meiotically heritable and that do not entail a change in DNA sequence.

[11] Using for example, hierarchical indexing for spliced alignment of transcripts (HISAT2)

[12] Rac guanine nucleotide exchange factor

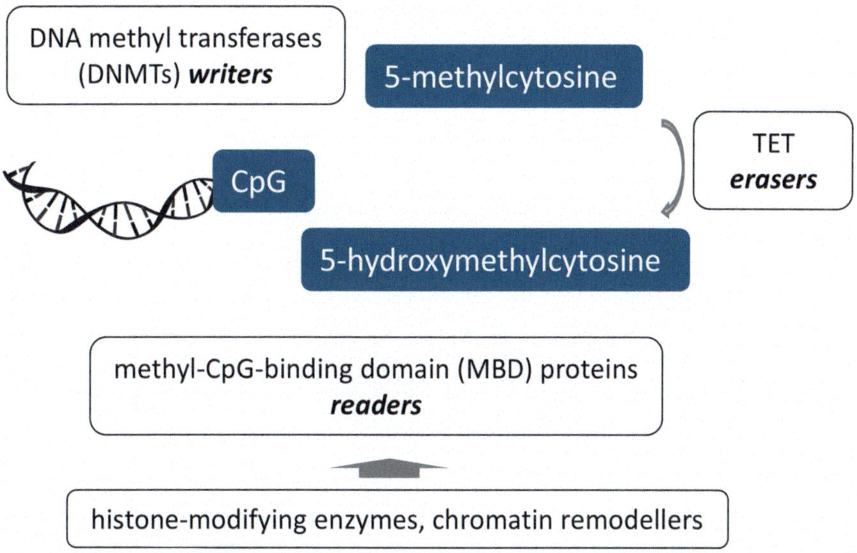

Fig. 5.8 DNA methylation

Epigenetic modifications (*marks*) consist of histone variants, posttranslational modifications of amino acids on the amino-terminal tail of histones, and covalent modifications of DNA and RNA bases. All contribute to the regulation of gene expression (i.e. switching on or off) during different developmental phases such as embryogenesis or in diseases such as cancer.[13] Epigenetic mechanisms have potential for the discovery of novel drug targets and are covered below based on a review by Yang et al. (2023).

DNA Methylation

Figure 5.8 illustrates the main features of methylation cytosine by DNA methyltransferases (DNMTs) to form 5-methylcytosine at predominantly CpG motifs in genomic DNA. Interpretation, addition and removal of epigenetic marks is likened to a process of *reading*, *writing* and *erasing*. Methylation writes marks onto DNA which allows reading by methyl-CpG-binding domain proteins to recruit histone modifying and chromatin remodelling enzymes for downstream functional effects. 5-methylcytosine can be converted to 5-hydroxymethylcytosine (and other derivatives) by a series of TET enzymes (erasers).

Methylation at the single base level in genomic DNA is usually determined using *bisulfite sequencing* which specifies the *DNA methylome*. Bisulfite treatment deaminates cytosine to uracil which is then read as thymidine after preparation for next generation sequencing (NGS). Methylated cytosines are unaffected and remain

[13] As well as the fact that monozygotic twins may not be truly identical at the phenotypic level

as cytosines, so treated and untreated DNA sequences may be compared to locate the position of epigenetic cytosine methylations. In keeping with the other genomics techniques mentioned in this chapter, methylation patterns can also be detected in single cells. This has been reviewed by Liu et al. (2023) who also provide some background to the role of DNA methylation in normal development and diseases such as cancer.

Histone Modifications

Core histone proteins associate with DNA to form nucleosome structures that enable about one metre of linear DNA to be packaged into the volume of the cell nucleus. The histones are also important regulatory molecules whose function is modulated through a series of post-translational modifications at the N-terminal *tails*. The most common modifications are shown in Fig. 5.9 along with a schematic diagram of the histone octamers that form the nucleosome on DNA.

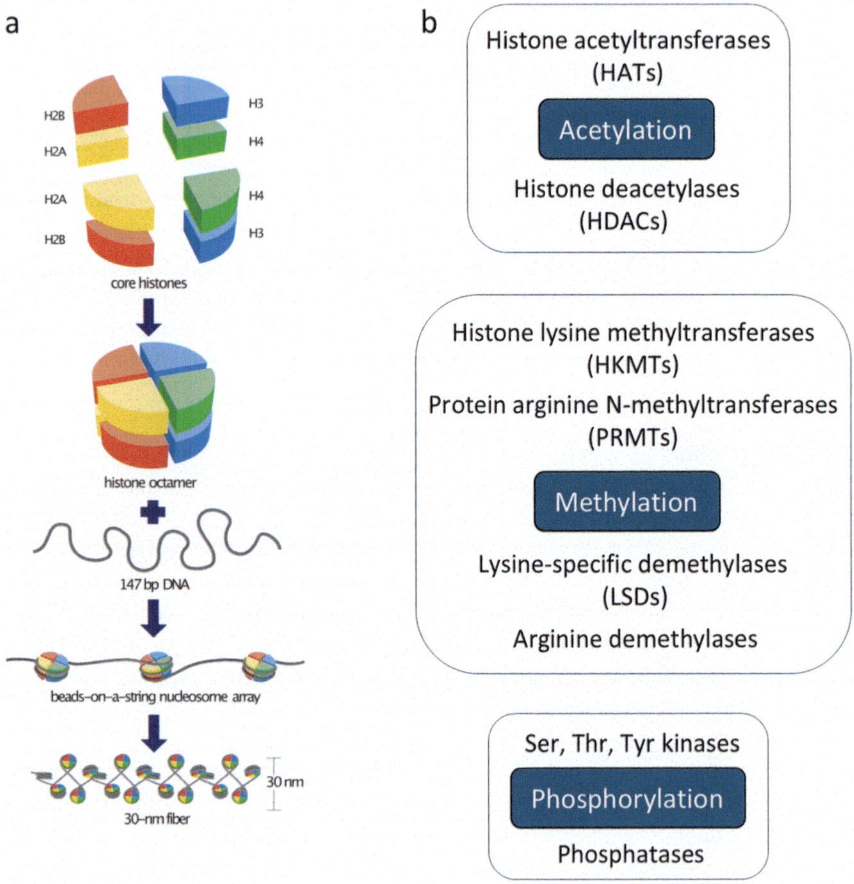

Fig. 5.9 Main histone modifications. (**a**) Nucleosome layout with histone octamer from Morgan (2007) under CC-BY-4.0 (**b**) Main post-translational modifications at the N-terminal *tail*

Table 5.7 RNA modifications

Modification	Location	Writer	Eraser	Reader
N6-methyladenosine	3′UTRs	m^6A methyltransferases	m^6A demethylases	m^6A-binding proteins
N1-methyladenosine	5′UTRs, tRNAs	TRM6–TRM61, TRM10	ALKBH1 ALKBH3	YTHDF3
5-methyl cytosine	Many RNA types	DNMT2, NSUN1–7	m^5C demethylases	ALYREF, YBX1
N7-methylguanosine	Many RNA types	METTL1/WDR4	m^7G demethylases	QKI5–17

As with DNA methylation, the enzymes involved in the addition or removal of acetyl, methyl and phosphate groups, writers and erasers respectively, modulate histone function in a variety of situations. The modified histones can interact with readers to associate functional assemblies with chromatin, thereby regulating gene expression. A prominent example is the BET family of proteins that interact with acetylated histones and are potential drug targets in cancer (see review by Yun et al. 2011).

Other histone modifications include ADP-ribosylation, deamination, O-GlcNAcylation, ubiquitination and lactylation, the last connecting metabolism to chromatin activation. The variety of histone modifications highlights the diversity of regulatory pathways acting on transcription in the cell nucleus (Zhang et al. 2019).

RNA Modifications

All the known types of RNA are subject to epigenetic modification, notably methylation. Table 5.7 lists the modified nucleotides, location, or types of RNA and the proteins acting as writers, erasers and readers. The field is evolving, so new candidates will appear over time, perhaps acting only at specific RNAs such as tRNA.

Other RNA modifications listed by Yang et al. (2023) include:

Adenosine-to-Inosine RNA Editing

Adenosine deaminase acting on RNA (ADAR1–3) deaminates adenosine to inosine on double stranded RNAs (dsRNAs) and can recode protein sequences if occurring in coding regions (I is mistaken for G); however, most occur in noncoding regions to change splicing patterns.

Pseudouridine (Ψ)

This C5-glycoside isomer of uridine is one of the most abundant RNA modifications and found in most types of RNAs.[14] Pseudouridine synthase (Ψ synthases PUS family) is the writer enzyme converting uridine to pseudouridine in an RNA-independent mechanism. One clear role for the modified nucleotide is in

[14]It came to prominence as a key component of the mRNA vaccines used for COVID-19

stabilising tRNA structures and other small RNA species and possibly modifying RNA splicing. Otherwise the area is still under active investigation.

N4-Acetylcytosine

Acetylation of cytosine at the N4 position is the only acetylation event known to occur in eukaryotic RNA. N-acetyltransferase 10 (NAT10) is considered to be the only writer enzyme, which in conjunction with U13 and the RNA adapter protein THUMP domain containing 1 (THUMPD1), acetylate rRNA and tRNA respectively. Cytosine acetylation controls pre-rRNA processing and ribosome production for 18 S rRNA and enhances tRNA and mRNA stability thereby promoting protein translation.

Chromatin Remodelling

Mammalian chromatin remodelling falls into four broad families according to their biological function and constituent proteins: SWI/SNF, imitation SWI, NuRD/MCHD and INO80. To achieve an active chromatin state, SWI/SNF complexes accelerate the ejection and insertion of histone octamers and facilitate the sliding movement of nucleosomes in an ATP-dependent reaction. ISWI family remodellers regulate the distances between nucleosomes, aid in chromatin assembly and transcription and randomize spacing, which improves RNA polymerase II (RNAPII) activation. INO80 can increase transcriptional activation and DNA repair.

Noncoding RNAs

Small noncoding RNAs (sncRNAs) include small nucleolar RNAs (snoRNAs), microRNAs (miRNAs),[15] small interfering RNAs (siRNAs), PIWIinteracting RNAs (piRNAs), extracellular RNAs (exRNAs), and circular RNAs (circRNAs). These, and the long noncoding RNAs (lncRNAs) are all implicated in cancer through regulation of transcription via epigenetic mechanisms.

Epigenetics and Drug Development

It will be clear from the above sections that there are many possible epigenetic modifiers that could be suitable as drug targets particularly since many are enzymes and therefore amenable to small molecule inhibition and even include the protein-protein interactions between readers and marked histone targets. Marketed drugs include DNA methyltransferase (DNMT) inhibitors such as azacitidine and decitabine and histone deacetylase (HDAC) inhibitors such as vorinostat and belinostat. However, they act on global transcription and therefore present risks of unacceptable toxicities. Of course there is a great deal of early development activity to focus drug therapies to more localised targets within the genome. An example is the RNA methyltransferase METTL3 which is implicated in oncogenesis and where an inhibitor developed by Storm Therapeutics has entered phase I trials (Storm Therapeutics 2025). The various epigenetic enzymes do not act in isolation but are

[15] The 2024 Nobel Prize in Physiology and Medicine awarded for their discovery

associated with other proteins that can modulate their activity; there are plenty of target opportunities even if they are not always suitable for intervention with small molecules.

5.3.1.6 Genome Editing

One of the most powerful ways of determining the function of an unknown gene is to selectively remove its sequence and observe any resulting phenotypic changes. Genome editing performs this function by removal of DNA at levels from single base pairs through to larger inserts; sequences can also be introduced in a selective manner. The therapeutic applications of DNA editing will be covered in Chap. 6, but the background science and future opportunities are presented here.

Background

As with so much in modern biotechnology, genome editing methods arose from the properties of bacteria and their bacteriophage invaders. The *arms race* between the two is well documented (Hampton et al. 2020) giving rise to a large variety of defence systems such as the restriction-modification system which produced the restriction enzymes that are essential tools for genetic engineering. Bacterial genes for the many identified anti-phage defence systems tend to be clustered in so-called genomic *defence islands* thereby providing an opportunity to identify novel components for future study (Doron et al. 2018). Zheng et al. (2024) have reviewed different genome editing tools, starting with earlier approaches such as zinc finger nucleases (ZFNs) and TALENS. The most high-profile editing system arising from the anti-phage systems is CRISPR-Cas9 which has revolutionised gene editing through its simple high-throughput implementation. Like any technology CRISPR-Cas9 has its limitations, which is why there are ongoing efforts to optimise the system for different scenarios and it is reasonable to expect that completely new editing techniques, perhaps based on other components of the *defence islands* will be developed in the future.

CRISPR-Cas9

Like higher organisms, bacteria show adaptive immunity, in this case by *remembering* prior infection by bacteriophages and resisting reinfection through the CRISPR-Cas9 system. The different components of the system were identified from the late1990s onwards and developed into a powerful gene editing method in the early years of the twenty-first century. An interesting background history is provided by Eric Lander (Lander 2016) from where some of the following is taken. *Clustered Regularly Interspaced Palindromic Repeats* (CRISPR) were detected in DNA from archaea and bacteria in the form of multiple copies of a near-perfect, roughly palindromic, repeated sequence of 30 bases, separated by spacers of roughly 36 bases. After much speculation as to the role of CRISPR it became clear that bacteriophage and invading plasmids both contained spacer sequences. The industrial production of yoghurt became the unlikely source of further hints of CRISPR's function. Spoiling of *Streptococcus thermophilus* used in dairy cultures were a problem for Rhodia Foods in France who used DNA analysis to determine how

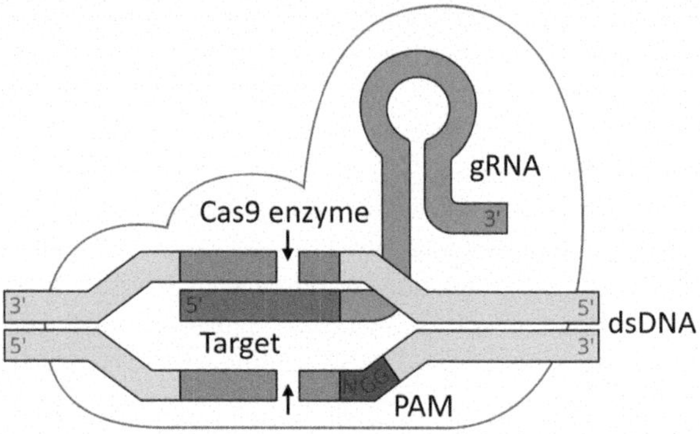

Fig. 5.10 Basic layout of CRISPR-Cas9 system. (Figure courtesy of Marius Walter under CC-BY-4.0). Double stranded target DNA (dsDNA) is hybridised with a guide RNA (gRNA) containing the sequence to be altered. A protospacer adjacent motif (PAM) is present near the target sequence and the whole arrangement allows for association with the Cas 9 enzyme and subsequent dsDNA cleavage

some bacterial strains became resistant to the causative phage infections. Resistant bacteria were found to have multiple copies of phage spacer sequences whose number correlated with increased resistance. Genes encoding *Cas* (CRISPR-associated) enzymes were also shown to be involved in resistance, with Cas9 being an active component of the bacterial immune system. Further details of how the CRISPR-Cas9 system worked became available over the years, including its implementation as a widely applicable gene editing tool, earning the discoverers Emmanuelle Charpentier and Jennifer Doudna the 2020 Nobel Prize in Chemistry. Double stranded DNA breaks are created by introducing to cells a guide RNA (gRNA) and the Cas9 enzyme. The gRNA contains a complementary sequence to the desired DNA target along with sequences facilitating association with the Cas 9 enzyme. Genomic DNA contains multiple random PAM (protospacer adjacent motifs) that must be linked to the target sequence to allow Cas 9 to cleave the double stranded DNA at the selected target sequence. After cleavage, the cellular DNA repair system fills the gap but disrupts the sequence. It is also possible to use a different DNA repair system to insert a defined sequence (Fig. 5.10).[16]

The Cas Enzyme Family

Cas 9 is one member of a much larger protein family some of which have even been identified in the most primitive archaeal organisms. Different Cas proteins have different properties and have great potential for fine-tuning the editing of RNA as well as DNA. The various subtypes of CRISPR-Cas have been reviewed by Makarova et al. (2020). Two main classes exist with different Cas enzymes namely:

[16] DNA repair will be covered later in this chapter

Class 1: Types I II IV with combinations of Cas 3, 8 or 10,11, 7, 5, 6, 1, 2, 4 linked to CRISPR loci interspersed with spacers.
Class 2: Types III V VI with combinations of Cas 9,12,or 13,1, 2, 4 linked to CRISPR loci interspersed with spacers.

In order to be reliable for drug target evaluation and therapeutic applications, DNA editing systems must be highly efficient through eliminating off-target effects and cellular stress responses as well as minimising the size of the CRISPR-Cas complexes for delivery. The cDNA for Cas9 is approximately 4.2 kb, which is a problem when using adeno-associated virus (AAV) vectors with an insert size limited to less than 4.7 kb. Novel Cas variants have been produced to overcome this size restriction including the smaller Cas12f construct. Inevitably, machine learning is playing a part, in this case by identifying Cas variants from large training sets of DNA and protein sequences as well as RNAs for guide RNA (summarised by Callaway 2024a, b). Ruffolo et al. (2024) identified over 1 million CRISPR operons by mining 26 terabases of assembled genomes and meta-genomes with some the resulting candidate gene editors (along with guide RNAs) comparable or even better in terms of accuracy than the Cas9 used in standard gene editing. Their best candidate was named OpenCRISPR-1 and could also be used for base editing (see below).

Base Editing

Gene editing using CRISPR-Cas introduces double stranded breaks in DNA which are reformed using *homology directed repair* (HDR), a not always ideal situation. Base editing removes the need for this repair by introducing point mutations in target DNA without cleavage. This is achieved by using a CRISPR-Cas construct where the Cas enzyme is engineered to be catalytically inactive (*dead-Cas*) and therefore unable to cut DNA. The Cas protein is linked to a ssDNA deaminase for different base conversions[17] as follows: cytosine C•G-to-T•A base pairs; adenine A•T-to-G•C base pairs; cytosine transversions (C•G-to- G•C base pairs (Zheng et al. 2024).

Prime Editing

Prime editing can produce 12 combinations of single or multi-base substitutions as opposed to six with base editing. Prime editors are formed by combining a deactivated nuclease domain in Cas9 with a reverse transcriptase (MMLV-RT) domain. A pegRNA template (generated from the Cas9 spacer sequence and a reverse transcriptase template containing the desired modification and primer binding site) is used to direct the Prime editor protein to its target. This is explained in more detail in Zheng et al. (2024), but the end result is precise single base editing that can be achieved in non-dividing cells, unlike with systems using homology directed recombination.

[17] Along with relevant DNA repair enzymes to complete the base conversion

Epigenome Editing

Dead-Cas proteins can in principle be fused to any protein domain to bring it into proximity with the DNA target of interest, as with base and prime editing. It is also possible to introduce epigenetic modifications such as methylation and demethylation or introduce d-Cas binding at two different sites in chromatin and bring them into proximity by crosslinking, thereby mimicking a mechanism for the regulation of gene expression *in vivo*. These and other modifications are described in more detail by Brezgin et al. (2019).

RNA Editing

DNA editing is clearly useful for target discovery and therapeutic use, but it may be desirable to make changes in RNA instead. This has the advantage of altering cellular functions without permanently altering the genome. The type VI CRISPR-guided endoribonuclease Cas13 has been used to remove RNA sequences but will activate bystander nucleases that degrade off target RNAs. An alternative method has been developed by Nemudrala et al. (2024) who used type III-A CRISPR complexes that cleave RNA in six nucleotide increments. The resulting RNA fragments are repaired, allowing for programmable deletions in a similar way to DNA.

RNA Interference (RNAi)

RNA-mediated interference (RNAi) silences gene expression through degradation of the double-stranded RNA formed between *small interfering RNAs* (siRNAs) and a homologous copy of cellular mRNA. The phenomenon was first noted as an antiviral defence system in plants but has now been used in mammalian cells and developed into RNA therapeutics (covered further in Chap. 6).

The basic mechanism is outlined in Fig. 5.11. Double stranded RNA or hairpin RNA is cleaved by the DICER enzyme to form small interfering RNA which activates the RNA-induced silencing complex (RISC). The complex unwinds the siRNA, releasing one strand (passenger) while holding the other to attract a complementary mRNA sequence. The resulting hybrid is cleaved by the RISC complex thus selectively preventing translation of the mRNA into protein.

The RNAi system has been to identify drug targets through disruption of mRNA expression (some examples reviewed by Masanori et al. 2005). Like CRISPR-Cas complexes, siRNAs can be incorporated into libraries to modify multiple genes or RNA expression in a single experiment.

Anti-CRISPR Proteins

Given the evolutionary arms race between bacteria and their phages it is perhaps unsurprising that the latter have developed systems for countering CRISPR action. A large group (over 90 families) of anti-CRISPR proteins (Acrs) have been identified with diverse functions (Trost et al. 2024). Acr proteins are interesting because they are able to disrupt large ribonucleoprotein complexes with one, AcrIF25, being able to achieve this without energy input from ATP hydrolysis. This property could be

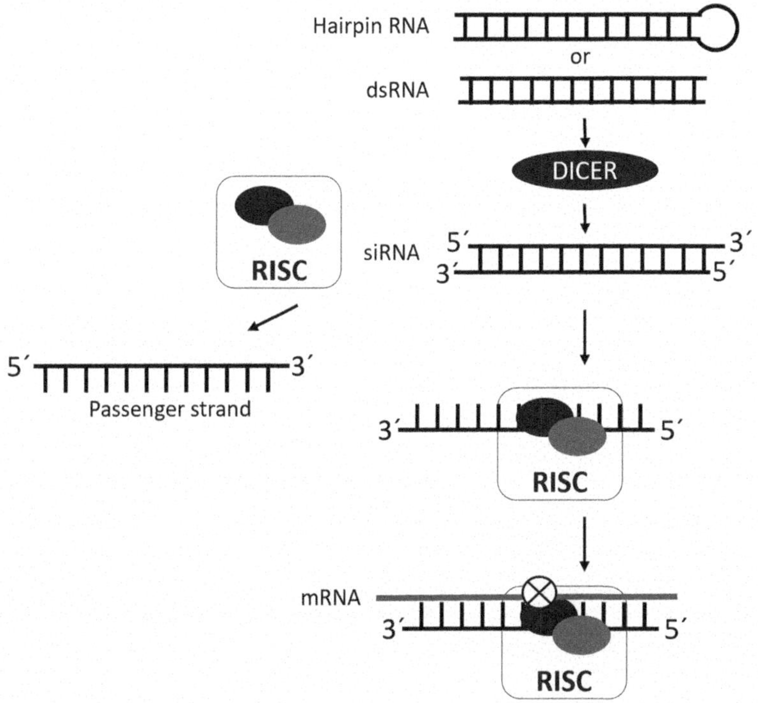

Fig. 5.11 RNA interference mechanism

exploited to disrupt other complexes and according to Trost et al. (2024), also provide a novel means of packaging RNA for delivery, clearly something to bear in mind for mRNA vaccination strategies.

Gene Editing Beyond CRISPR-Cas

It is clear from the above that a detailed understanding of the strategies employed by archaea and bacteria to block bacteriophage and plasmid infection (and their countermeasures) has provided powerful gene editing tools based mostly on the CRISPR-Cas system. However, there are other mechanisms under investigation that could be used not only gene editing, but also for biotechnological applications not necessarily involving changes to DNA. Two examples follow here, with references to provide detailed background.

DNA-Editing Transposases

Mobile genetic elements (MGEs) contribute to the genome diversity that drives evolution and mobilise through the action of various enzymes, including transposases, integrases, endonucleases and recombinases. Insertion sequence (IS) elements are found in archaea and bacteria with IS110 elements integrating in in microbial genomes in a sequence-specific manner. Durrant et al. (2024) show that a circular form of IS110 drives the expression of a non-coding RNA with two distinct binding loops that separately recognize the IS110 DNA donor and its genomic insertion target site. These sites can be bridged by base pairing with programmable RNA sequences allowing the IS110 recombinase to alter diverse target sequences (summarised by Ledford 2024). Although the IS110 (and the similar IS111) systems have not yet been shown to work in mammalian cells, they could have utility if they do, since the RNA and protein involved is smaller than CRISPR-Cas and therefore easier to package into vectors.

DdmDE Anti-plasmid System

The DNA defence module DdmDE in *Vibrio cholerae* rapidly eliminates small multicopy plasmids (7PET) using a combination of DNA targeting and recruitment of a protein complex with nuclease activity (Bravo et al. 2024). DdmE is a catalytically inactive, DNA-guided, DNA-targeting prokaryotic Argonaute[18] variant (eukaryotic Argonaute proteins (eAgos) use RNA interference (Swarts et al. 2014)). DdmE uses a 5′-monophosphophorylated DNA guide to selectively target complementary DNA sequences and load a single-stranded DNA target via DdmD into the RecA helicase channel culminating in activation of the complex. This triggers the disassembly of the DdmD dimer to facilitate processive DNA degradation (detailed in Bravo et al. 2024). The causative agent of cholera, *Vibrio cholerae*, causes millions of deaths each year, so genetic engineering strategies based on the DdmDE system could be developed to create anti-infective agents as well as novel gene editing tools.

Reverse Transcription

Reverse transcriptases (RTs) are used by bacteria in antiphage defence systems including CRISPR-Cas and *retrons*. The latter consist of a combination of RT, ncRNA, and an effector protein toxin (Osterman and Sorek 2024).These proteins are produced by a convoluted process in which bacterial DNA is transcribed to form a noncoding RNA which is converted back to DNA by a rolling circle reverse transcription reaction; the resulting DNA is transcribed into mRNA which is translated into the toxic protein. The possible advantages to bacteria of having this unusual protein synthesis mechanism include control of bystander toxicity, autoimmunity and lytic phage infection (Osterman and Sorek 2024). For biotechnology applications, bacterial retrons are being used for genome editing because of their ability to endogenously generate a cDNA of choice thus overcoming issues with the delivery of exogenous DNA templates (Lopez et al. 2022).

5.3.1.7 Synthetic Lethality

Genes do not operate in isolation but interact in networks and co-dependencies that can provide new drug targets and therapeutic options. *Synthetic lethality* (SL) is the

[18] A mutant gene was originally identified in plants whose leaves curled up like squid tentacles leading to the name taken from that of the octopus *Argonauta argo*

Table 5.8 DNA repair pathways

MMR	Mismatch repair
BER	Base excision repair
NER	Nucleotide excision repair
TLS	Trans lesion synthesis
HR	Homologous recombination
NHEJ	Non homologous end joining
FA	Fanconi anaemia
Reversal of DNA lesions	Removal of O^6 alkylguanine

combined alteration of individually viable genes, most activity being in oncology where many somatic genetic changes occur and involve DNA repair pathways (reviewed in Schäffer et al. 2024). Previtali et al. (2024) provide a description of how SL is being applied to novel drug target discovery. Possible scenarios:

- Alteration in gene A justifies targeting gene B
- Developing synergistic drug combinations e.g. pair of genes (A and B)
- Synergistic drug combinations that exploit a mutation in a third gene (C)
- Stratifying cancer patients for tailored therapies.

The SL approach has the potential to identify actionable targets that are linked to clinically relevant loss of function or undruggable targets that are hard to affect directly.

DNA Repair Pathways

Many thousands of DNA lesions occur daily in each cell and must be corrected by the *DNA damage response* (DDR) system to avoid compromising the host organism.

The following section focuses on the DNA repair and cell-cycle checkpoint pathways activated by the DDR since these have provided the majority of SL targets. The major pathways are reviewed in Jackson and Bartek (2009) and listed in Table 5.8.

PARP Inhibitors- First-Generation SL Drugs

Tumours with BRCA1 or BRCA2 mutations are deficient in homologous recombination (HR) and unable to repair double stranded DNA lesions. The poly (ADP-ribose) polymerase 1 and 2 (PARP1,2) proteins repair single-stranded breaks but are vulnerable to inhibition thereby forcing double stranded breaks (DBBs) in the DNA. In HR deficient tumours with the BRCA mutation, for example, treatment with the PARP inhibitor olaparib leads to an increase in DSBs and resulting cell death. Since normal cells lack the BRCA mutation, PARP inhibitors have become valuable therapies against breast and ovarian cancers Previtali et al. (2024). One of the proposed mechanisms of action of PARP inhibitors involves trapping the PARP enzyme on chromatin, but the details may be more nuanced than initially suggested (Petropoulos et al. 2024). A different use of PARP inhibitors has been described in a clinical trial for glioblastoma where PARP is over expressed and can be targeted with a small molecule labelled with ^{123}I, a radioisotope with a localised emission of low energy electrons used to spare bystander tissue (Ariceum n.d.).

Other SL targets related to PARP include RAD51 and 52 and Poly(ADP-ribose) glycohydrolase (PARG) an enzyme that is often upregulated in cancer and positively correlated with resistance to PARP inhibition. DNA polymerase θ (Polθ, POLQ) is a high-profile SL target. The enzyme is part of the TMEJ pathway that provides a backup to NHEJ or HR for dsDNA repair if the latter are compromised in cancer. Polθ is overexpressed in colorectal, breast, ovarian, lung, and gastric cancer but at low to zero levels in normal cells.

Other SL targets relating to the DDR have been identified, in this case the mismatch repair (MMR) system, deficient in a number of solid tumours and leading to *microsatellite instability* (MSI) though insertion/deletion events in repetitive DNA tracts. The Werner syndrome RecQ helicase was identified in the large scale screens *project Achilles* and *project DRIVE*; the former using CRISPR–Cas9 on 517 cell lines and the latter, RNA interference on 398 cell lines to define genes essential for proliferation and survival of individual cancer cell lines (Chan et al. 2019). WRN was found to be the top dependency with an inhibitor that shows synthetic lethality in cancer (Ferretti et al. 2024). Mismatch Repair also has a role in other conditions. For example, the fatal neurodegenerative disorder Huntington's Disease is characterised by significant expansion of the CAG trinucleotide repeat in exon 1 of the (*Huntingtin* locus caused by a dysfunctional MMR system. In this case inhibitors targeting two key components MutSβ and MutSα may have therapeutic utility (e.g. LoQus23 2025).

Computational Approaches to SL Target Discovery

Laboratory investigations of potential synthetic lethality (SL) gene combinations are clearly effective, but also limited since the latter represent a tiny minority of the theoretical possibilities. This is where computational methods, including machine learning and deep learning methods can play a role (Wang et al. 2022). These authors list some databases of SL pairs, including SynLethDB which is curated from lab-based, literature and bioinformatics resources such as DAISY that employ statistical inferences when mining cancer genomics data for SL gene pairs. Firstly, genomic survival of the fittest (SoF) is based on the observation that cancer cells are strongly selected against if they have lost two SL-paired genes; secondly, the knock down of a synthetically lethal gene is lethal to cancer cells where its SL partner is inactive; thirdly, SL pairs tend to participate in closely related biological processes and hence are likely to be coexpressed. Details about these procedures and the development of DAISY are given in Jerby-Arnon et al. (2014).

The methods reviewed by Wang et al. (2022) are based on searches for pairwise gene interactions, but gene networks are more complex in biological systems. The scale of the problem can be illustrated by an analysis of yeast trigenic interactions. Out of ~200,000 cells, 3196 trigenic interactions have been detected, representing a 100-fold increase over two-gene interactions (quoted by Wang et al. 2022).

SL research is a very promising approach for identifying novel drug targets (e.g. PARP inhibitors) and while still at an early stage, shows great promise for future drug discovery.

5.3.1.8 Synthetic Rescue

The complex interaction between genes and phenotypes is a recurrent theme in genomics research as illustrated with synthetic lethality. However, mutations can confer a protective effect through *genetic suppression* by reducing the impact of a deleterious mutation. Mechanisms of suppression have been described in model organisms and may exist in humans, thereby explaining observed variations in disease severity or penetrance. In effect, mutations that put an individual at high risk of developing a disease can be overridden by mutations elsewhere in the genome.

This concept of *synthetic rescue* (SR) is being investigated commercially by Adrestia (now acquired by Insmed) who recognise the opportunities for drug target discovery (Eisenstein 2024; Insmed). A detailed background to current SR target discovery efforts is given by Ünlü et al. (2023) who note that protective modifiers may exist for most common monogenic genetic diseases, including sickle cell disease, β-thalassemia, cystic fibrosis, Huntington's disease, Duchenne's muscular dystrophy, and spinal muscular atrophy. Lastly, the deubiquitylating enzyme USP48 is synthetically viable for mutations in the Fanconi anaemia (FA) DNA repair pathway (Velimezi et al. 2018). FA is a rare disease giving an increased cancer risk due to defective repair of DNA through FA-gene deficiencies. Compared to FA-defective cells alone, FA-deficient cells additionally lacking USP48 are less sensitive to genotoxic stress induced by ICL agents and display enhanced, BRCA1-dependent, clearance of DNA damage. USP48 inactivation reduces chromosomal instability of FA-defective cells and therefore could be an actionable synthetic rescue target.

5.3.2 Proteomics

The significant advances in large-scale detection and analysis of nucleic acids provide the tools for analysing the molecular basis of disease at the level of mutated or epigenetically modified DNA and the expression of coding and regulatory RNAs. However, the drug targets revealed by these approaches are mostly proteins and therefore proteomics is fundamental to drug target selection.[19] Figure 5.12 shows the workflow needed to identify and analyse proteins as potential drug targets.

5.3.2.1 Primary Sequence Determination

Technologies for generating primary amino acid sequences of proteins and peptides have come a long way since the pioneering methods developed by Sanger and later,

[19] As well as the therapeutic proteins covered in the next chapter

Fig. 5.12 Proteomic hierarchy relating to drug targets

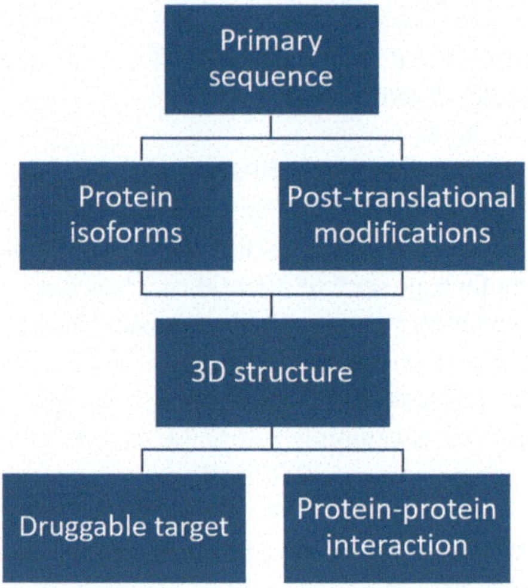

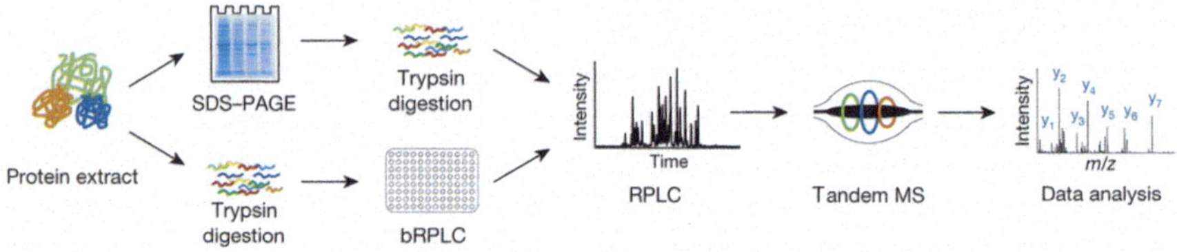

Fig. 5.13 Proteome analysis using tandem mass spectrometry. (Figure from Kim et al. 2014 with permission)

Edman. High resolution tandem mass spectrometry is now routinely used to analyse sequences in protein mixtures derived from cells and other biological material (*shotgun proteomics*) and has been applied to the whole human proteome as described by Kim et al. (2014). At that time the human genome sequence contained approximately 20,687 protein-coding genes of which 17,294 (about 84%) were identified by proteomic profiling of 30 normal human tissues and primary cells. The analysis protocol described by Kim et al. (2014) is shown in Fig. 5.13. Proteins in cell or tissue extracts are unfractionated or separated using SDS-PAGE prior to digestion with trypsin. The resulting tryptic peptides are separated using reverse phase liquid chromatography (RPLC) and subjected to tandem mass spectrometry. The resulting mass to charge ratios (*m/z*) are compared with a *virtual tryptic digest* of the human proteome in order to relate experimentally derived peptides to protein sequences using MASCOT and SEQUEST bioinformatic software.

The methods described here are a huge advance on earlier low throughput techniques and provide primary sequence data at scale. However, it is not straightforward to identify the post-translational modifications, products of alternative splicing and single amino acid polymorphisms that are involved in protein function and that substantially expand the repertoire of proteins beyond the number of encoding genes. Sinitcyn et al. (2023) have had some success in this area by using multiple protease digestions in addition to trypsin which provides a more diverse peptide repertoire for analysis. Mass spectrometry methods are clearly useful, but in an ideal world it would be desirable to have a protein sequencing technology with the same throughput and flexibility as that used with DNA and RNA. Unfortunately, unlike nucleotides, the amino acids that form proteins have significant variation in charge and mass. However, based on proof-of-concept experiments, there is some hope that adapting DNA nanopore sequencing to proteins may be feasible. The ideal situation would be to sequence an entire polypeptide chain and detect any post-translational modifications if present. To achieve this, the protein must be unfolded and passed through a nanopore slowly enough to register each amino acid as an electric current. Motone et al. (2024) have developed a system using proteins tagged with a negatively charged protein designed to reversibly draw the molecule through a nanopore using electrophoresis; a *stopper* group is added to the other end to retain the assembly in the nanopore. The modified protein is then drawn back through the nanopore using the ClpX unfoldase enzyme. The flow rate is such that amino acids can be read as an ion current in the MinION flow cell used for DNA sequencing. Although the system is in the early stage of development it points the way towards protein sequencing technology that compares with that for DNA in terms of ease of use and throughput.

Single Cell Proteomics

As with DNA and RNA analysis, single cell protein analysis using mass spectrometry (scp-MS) has the potential to identify mechanisms and drug targets that would be obscured in ostensibly homogeneous cell populations. To quote from a review by Petrosius and Schoof (2023):

Mass-spectrometry analysis of proteins generally requires >100 ng of input material. However, one mammalian cell is estimated to contain only 100–200 pg of protein. Unlike for genome sequencing-based methods, there is currently no known way to artificially amplify the amount of protein in a cell, making it essential to minimize sample loss during processing.

The latter problem can be overcome by using nanodroplets or FACS sorted cells in microwells. Refinements to the mass spectrometry instrumentation such as Field Asymmetric Ion Mobility Spectrometry (FAIMSPro) allows the elimination of singly charged contaminant ions and selection of the multiply charged peptide ions of interest. Scp-MS has been applied to different biological problems and is complementing RNA-based methods in a *multiomics* approach to studying single cells (see later).

5.3.2.2 Protein Isoforms and Post-translational Modifications

As mentioned earlier, the number of human proteins far exceeds the roughly 20,000 encoded by the human genome due to alternative splicing of mRNAs to produce protein isoforms as well as post-translational modifications (PTMs). Since these phenomena impact upon normal biological and pathogenic disease mechanisms there is a strong need for analytical techniques that are both sensitive and high

Table 5.9 Post-translational modifications

	Short chain fatty acids	Long chain fatty acids	Redox modifications
Phosphorylation	Propionylation	Palmitoylation	S-Nitrosylation
Acetylation	Butyrylation	Myristoylation	S-Glutathionylation
Methylation	Malonylation		Sulfenylation
Ubiquitination	Succinylation		Sulfinylation and sulfonylation
SUMOylation	2-Hydroxyisobutyrylation		Disulfides
Glycosylation	β-Hydroxybutyrylation		
Citrullination	Crotonylation		
Carbamylation	Glutarylation		
ADP-ribosylation	Lactylation		
Benzoylation			
Neddylation			

From: Zhong et al. (2023) under CC-BY-4.0

throughput. Different protein isoforms can be expressed in different tissues which creates the opportunity to target specific mechanisms. For example, pan-isoform myosin light chain kinase 1 (MLCK1) inhibitors are severely toxic owing to global MLCK1 inhibition, whereas an isoform-specific domain in the intestine-specific MLCK1isoform is not, thus providing a feasible drug target in gastrointestinal disease (Kjer-Hansen et al. 2024). These authors review the opportunities for novel target discovery with protein isoforms as well the potential for targeted drug delivery, neoantigens and diagnostics.

Perhaps the most well-known PTMs are phosphorylation and glycosylation. However there are hundreds of modifications that may affect biological function, perhaps through reversible signalling mechanisms or structural features. The PTMs listed in Table 5.9 are taken from a comprehensive review of the field by Zhong et al. (2023).

From the above list it is clear that there is plenty of scope for future work on the biological functions of PTMs and the detection of new ones. Mass spectrometry is successful in detecting PTMs, but often requires enrichment of the target protein, using antibodies for example. The *bottom-up* approach analyses the proteome as peptides derived by digestion of intact proteins, while the *top-down* approach uses intact proteins in conjunction with new generations of highly sensitive mass spectrometers (Chen et al. 2018). These experiments generate complex mass spectra that may be challenging to deconvolute, an area where machine learning has a role to play (e.g. McIlwain et al. 2020) and will no doubt be fully exploited in the future.

5.3.2.3 3D Structure Determination

The importance of the protein structures of drug targets has already been highlighted in Chap. 3 in relation to virtual screening. Generating these structures has been a major part of molecular biology since the 1950s and recent advances in deep learning have revolutionised the field. The landmarks in structural biology range from the development of X-ray crystallography, through nuclear magnetic resonance (NMR) to the more recent technique of cryoelectron microscopy (reviewed by Curry (2015). The 3D structures of thousands of proteins have been deposited in the Protein Data Bank (RCSB PDB 2025) and provide the material for developing computer-aided structural prediction.

Cryoelectron Microscopy

Electron microscopy (EM) of biological samples was developed in the 1940s and has successfully imaged cells, viruses and DNA. Although electron beams can easily surpass the resolution limit imposed for light microscopy, the technique is unable to breach the 3 Å resolution threshold below which atomic details can be resolved and modelled reliably. This is due to the heavy metal staining required for transmission EM, a problem solved by snap freezing unstained samples in liquid ethane prior to performing cryoelectron microscopy (CryoEM). Since the development of hardware for electron detection and analysis software around 2012–14, cryo-EM has been used to generate thousands of structures,[20] notably of protein complexes such as those found in ribosomes (Azinas and Carroni 2023). The success of CryoEM does not necessarily mean that it will supersede X-ray crystallography for the structure determination of individual proteins, but it will continue to prove useful for imaging membrane proteins and larger cellular structures (see Sect. 5.3.4.1).

Deep Learning and Structure Prediction

Homology modelling, where only primary sequence information is available to determine protein structure has been used for some years. This has been successful where a high degree of homology exists between the unknown and reference structures, but excludes the majority of proteins of interest. The application of deep learning (see Chap. 3) to the problem has resulted in a transformation of the field though the development of AlfaFold for structure prediction (Jumper et al. 2021) and RoseTTAfold for protein design (Baek et al. 2021).[21]

The AlfaFold protein structure database (2025) contains over 200 million different structures from across many biological taxa which indicates the ease with which proteomes can be surveyed. The question for drug discovery scientists is whether the structures are sufficiently accurate and can also accommodate other molecules. The latest iteration, AlfaFold 3, can model protein interactions between proteins, nucleic

[20] For example, the Electron Microscopy Data Bank (EMDB) contains over 40,000 structures (The wwPDB Consortium 2024)

[21] The 2024 chemistry Nobel prize was awarded to Jumper and Hassabis (AlfaFold) and Baker (RoseTTAfold)

acids, small molecules, ions and modified residues (Abramson et al. 2024), and similarly RoseTTAFold All-Atom (Krishna et al. 2024). Inevitably, questions have been raised about the utility of the structure predictions in drug discovery. From reports described in a feature by Callaway (2024a, b), some of the AlfaFold predictions were unusable for standard virtual screening but a reasonable number were useful for jump starting a project thereby making significant time savings. While these problems are fully recognised, the future of protein structure determination will most likely be dominated by refinements of these deep learning models and training with an increasing number of experimentally derived structures. Structural data on protein domains generated by AlfaFold or equivalent will be useful for analysing drug target classes across the human and other genomes. As an illustration, Lau et al. (2024) have compiled an Encyclopedia of Domains containing nearly 365 million domains across more than 1 million taxa, revealing thousands of putative new folds for further investigation.

5.3.2.4 Protein Target Druggablilty

This last section of the proteomics flowchart in Fig. 5.12 covers the classification of protein into druggable and non-druggable targets. The rather inelegant term *druggable target* refers to proteins (or other macromolecules) whose activity can be modified with chemical molecules less than about 700 daltons in molecular weight. Molecules of this size are often orally bioavailable, straightforward to manufacture and easy to administer, hence their popularity with originators and patients alike. However, many interesting targets involve protein-protein interactions that are generally unsuitable for small molecule development due to low affinity interactions resulting from a lack of deep binding pockets. This problem has been recognised for decades and has held back the development of medicines for serious diseases using conventional medicinal chemistry. Despite heroic efforts to find suitable small molecules for the majority of these undruggable targets it eventually became clear that large biologicals such as antibodies or even cells were the only entities with the necessary potency and selectivity to be effective drugs in this situation. In a sense this realisation lies at the heart of modern drug discovery and development and forms the central theme of this book; drug development is encompassing these new approaches that are based on an increasing understanding of the immune system, genetics and cell biology. There are always exceptions to any rule however, and some supposedly undruggable targets have been affected by potent small molecules with clinical activity, something that is covered further in Chap. 6.

When surveying candidate targets for their druggablilty or otherwise, a range of virtual tools have been applied based on structural data. One example is an algorithm for classifying the depth of binding pockets (Kalidas and Chandra 2008) and there are many others. In addition to binding pockets on proteins, target druggablilty includes other features like biological function, subcellular location and interactions within a network. Machine learning has been used to analyse 20,404 human proteins based on the above parameters, including structures obtained using AlfaFold and binding pockets from the Fpocket database (Cunningham et al. 2023).

In addition to validating known druggable targets, the PINNED system developed by Cunningham et al. (2023) could identify putative druggable targets within the *dark proteome*, one example being the transmembrane serine protease TMPRSS11B. Interestingly, the dark proteome may also contain small *miniproteins* that have biological functions despite some examples being less than 12 amino acids long (Pennisi 2024).

Studies of this type add to the drug target identification workflow in a positive way, but with the caveat that the structural data produced by deep learning must be accurate to be useful.

5.3.3 Metabolomics and Metabonomics

To quote Lindon et al. (2003):

Proteomics, and genomics in particular, do not provide evidence of real-world endpoints for diagnosing disease or evaluating beneficial or adverse drug effects.

While the preceding sections of this chapter show that there have been major advances in these *omics* fields over the years since the Lindon et al. (2003) article was published, the point is still valid. Real world responses in tissues and cells involve measurable changes in their *metabolomes*, namely the composition of metabolites analogous to genomes and proteomes. *Metabonomics* takes this further by generating in addition to static concentrations of endogenous metabolites, full time courses of metabolite fluctuations measured in biological fluids such as blood, urine and cell culture supernatants using spectroscopic techniques such as NMR and mass spectrometry. While metabolite analysis is well suited for identifying biomarkers in disease and in toxicology, it also has a place in target discovery (Pan et al. 2024). Figure 5.14 illustrates how metabolite profiles can be used as fingerprints to classify cellular responses to drug treatments. Transcriptome profiles are often used with compound screening to group cellular responses into discrete target classes. Metabolomics can be applied in the same way, with the advantage that it measures these responses directly rather than through the intermediate level of mRNA expression (see also Garana and Graham 2022).
In keeping with the common theme running though this chapter, advances in cell metabolomics are being made at the single level along with the deployment of machine learning algorithms. Hu et al. (2023) present an example of this using a single cell spatially resolved metabolic (scSpaMet) framework incorporating untargeted spatial metabolomics and targeted multiplexed protein imaging in a single pipeline. This powerful technique was used to detect unique metabolic states within cell types, something that is particularly relevant to cancer where actionable metabolic targets are likely to be found. scSpaMet used mass spectrometry imaging techniques to delineate cell types using Imaging Mass Cytometry (IMC) and profile metabolites at the single cell level using ToF-SIMS imaging. The IMC method will be covered in the later section on cell biology. Time of Flight-Secondary Ion Mass

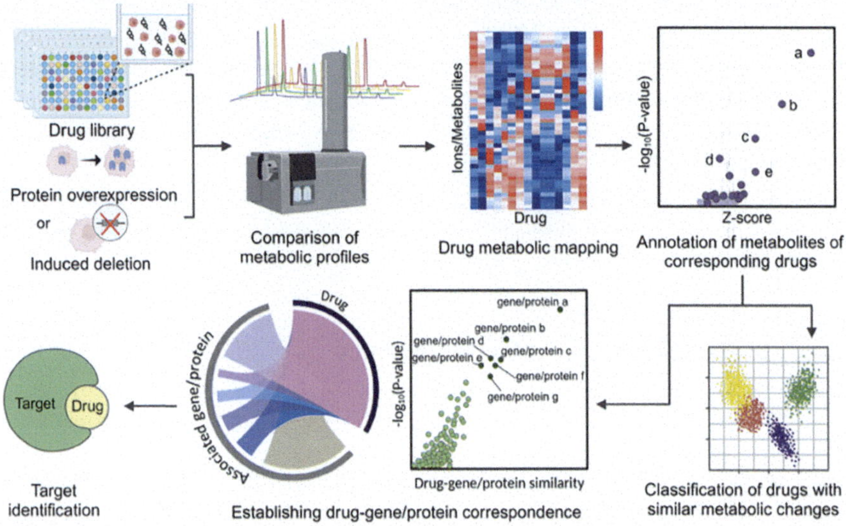

Fig. 5.14 Using metabolic profiles to classify drug responses. (Taken from Pan et al. 2024 under CC-BY-4.0)

Spectroscopy (ToF-SIMS) uses an ion source to generate secondary ions from the surface of the sample (a tissue section here) and to determine the m/z ratios of the latter using *Time of Flight* mass spectrometry. The highly sensitive procedure allows detection of molecules 1–2 nm below the surface. The Hu et al. (2023) paper provides some background and references.

5.3.4 Cell Biology

Understanding the structure and functions of cells is highly relevant to the study of disease pathogenesis and the search for potential drug targets. The following sections cover cell analysis methods and specific cellular functions like *proteostasis* and cell death mechanisms that have exploded as areas of study in recent years.

5.3.4.1 Analysing Cell Types and Functions

Microscopy

Light microscopes have been used for centuries to visualise cells and other structures, but their resolution is limited by the wavelength of light. The introduction of transmission and scanning electron microscopy in the twentieth century has enabled the visualisation of cells and organelles in unprecedented detail. However, the microscopy field never stands still and steady improvements in the basic technology, instrumentation and image analysis continue to be made in the present century. The title of a review by Nogales and Mahamid (2024) *Bridging structural and cell biology with cryo-electron microscopy* neatly summarises the need to increase the resolution of cell imaging down to the level of single molecules so that an entire cell can be seen for what it is, namely an assembly of interacting molecules. Although this is not yet achievable, improvements in cryoelectron microscopy (cryoEM) are moving the field in this direction. The development of cryoelectron tomography (cryoTM) allows 3D visualisation of cellular components *in situ* and can be used in synergy with cryoEM to study cell structures at high resolution. For example, structures of large macromolecular complexes, polymers or membrane proteins, as obtained by cryo-EM of purified complexes, can help to find those structures in the cell when seen in cryo-tomograms (Nogales and Mahamid 2024).

Figure 5.15 shows an example where Gemmer et al. (2023) have used cryoTM to visualise ribosomes bound to endoplasmic reticulum (ER) vesicles along with specific ribosome-associated proteins imaged at molecular resolution.

The high-resolution cryoEM and TM microscopy described above will be developed further and generate important data for basic biology and drug discovery. However, the procedures are costly and require dedicated facilities and trained personnel. There is still a place for light microscopy, not only for routine cell and tissue analysis, but also for more sophisticated applications. Fluorescence microscopy is long established, using molecules that fluoresce when bound to specific molecules (e.g. DNA) or using antibodies tagged with fluorophores. The fluorescence can be induced by conventional tungsten lamps, but confocal microscopy was

Fig. 5.15 CryoTM visualisation of different ribosome structures (top) to ER vesicles. (Adapted from Gemmer et al. 2023 under CC-BY-4.0)

Table 5.10 Some rare earth metal isotopes used in imaging mass cytometry

^{139}La	^{147}Sm	^{166}Er
^{141}Pr	^{151}Eu	^{169}Tm
^{142}Nd	^{152}Sm	^{170}Er
^{144}Nd	^{159}Tb	^{171}Yb
^{145}Nd	^{164}Dy	^{174}Yb
^{146}Nd	^{165}Ho	^{176}Yb

introduced using laser light to focus images from thick tissues. Further descriptions of light microscopy instrumentation and techniques can be found in many publications; one by Mertz (2023) covers advances in light microscopy, including expansion microscopy where samples are physically expanded to reveal nanoscale levels of detail; Consistent with the theme of this book, Balasubramanian et al. (2023) offer a future perspective on microscopy.

The imaging techniques described so far involve the detection of electrons or photons. Another option is to detect ions for analysis by mass spectrometry. Imaging Mass Cytometry (IMC) has already been briefly mentioned in the context of single cell metabolomics. Antibodies or other probes can be tagged with isotopes of rare earth metals allowing multiplexed detection by ToF mass spectrometry. Some examples of the metal tags are shown in Table 5.10.

Lastly, the imaging data for all microscopy techniques requires computational analysis, an area where machine learning plays a part. The publication by Carpenter (2023) *Smart microscopes of the future* lays out a vision where specimens can be imaged and specific features relayed to a human operator in natural language.

The Human Cell Atlas

Drug targets are, of course, present on the surface and interior of the cell, or are secreted into the external milieu. It follows therefore that a detailed understanding of the exact nature of cells comprising an organism, their interactions and roles in basic biology and disease is fundamental for target discovery. For obvious reasons, humans represent the primary organism for study, but clearly many others are relevant for infectious diseases and the actions of the microbiome. Adult humans are comprised of approximately 37.2 trillion cells with a multitude of different functions. Cataloguing the whole system in detail is an enormous undertaking that started in 2016 as the Human Cell Atlas (HCA) consortium (Human Cell Atlas 2025). As of 2024, the consortium has grown into a global network of more than 3600 members in 102 countries, contributing data to 18 Biological Networks. The HCA has now entered a phase of data integration towards the assembly of the first draft atlas and publication of early findings applied to specific biological situations. The experimental approaches used for this have already been outlined in this chapter and include single cell multiomics technology, imaging techniques and data analysis using machine learning and related computational methods (see Dance (2024) for a summary). Multiomics technology has the potential to detect the genome, epigenome, proteome and metabolome from the same cell. Not all of this has been realised at the time of writing (metabolomes are difficult to integrate with the others)

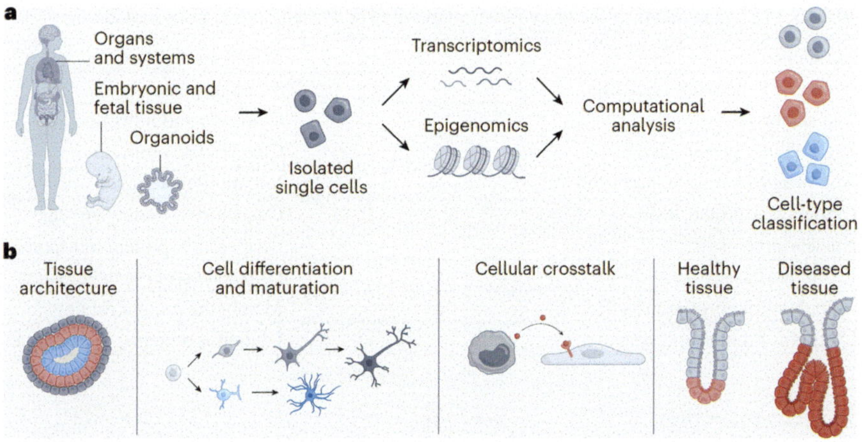

Fig. 5.16 Workflows of the Human Cell Atlas project. (Taken from Yanai et al. 2024 with permission)

but progress is being made, for example with the simultaneous analysis of cell cycle transcriptomes and proteomes from tumour cells (Mahdessian et al. 2021).

The key components of the HCA project, taken from Yanai et al. (2024) are shown in Fig. 5.16.

Other consortia for large-scale data generation related to cell functions include π-HuB, a collaboration for human proteomics analysis using advanced experimental and computational technology (He et al. 2024b). This, like the Human Cell Atlas consortium and related initiatives, should provide a large number of new opportunities for drug discovery and development over the coming decades.

5.3.4.2 Protein Homeostasis: Proteostasis

So far, this chapter has highlighted general approaches to drug target discovery with few specific examples. The following sections are different in that they cover a specific aspect of cell biology, namely the control of protein homeostasis or *proteostasis*. Other examples could be used, such as cell cycle control and cell signalling mechanisms, but addressing proteostasis in particular has become a very high-profile activity in the biopharma industry and one which shows great promise for the development of medicines with novel modes of action.

Proteins are fundamental cellular components that must be synthesised, translocated and degraded under the control of a network of components, the *proteostasis network*. These components, taken from the Proteostasis Consortium, are listed in Table 5.11. Some of their functions are relevant to neurodegenerative diseases such as Alzheimer's, Parkinson's, and amyotrophic lateral sclerosis that are characterized by the accumulation of misfolded proteins (Höhn et al. 2020). Some of the proteostasis mechanisms related to the categories listed in Table 5.11 are

Table 5.11 Components of the proteostasis network (The Proteostasis Consortium 2025)

Component	Selected functions
Proteostasis network regulation	Stress response
Translation	Cytosolic and mitochondrial
Extracellular proteostasis	Chaperones, proteases
Autophagy Lysosome pathway	Initiation, maturation
Cytonuclear proteostasis	Chaperones, folding enzyme
Nuclear proteostasis	Chaperones, protein transport
ER proteostasis	Organelle-specific degradation
Mitochondrial proteostasis	Membrane protein folding
Ubiquitin Proteosome system	Ubiquitin proteins, Proteosome

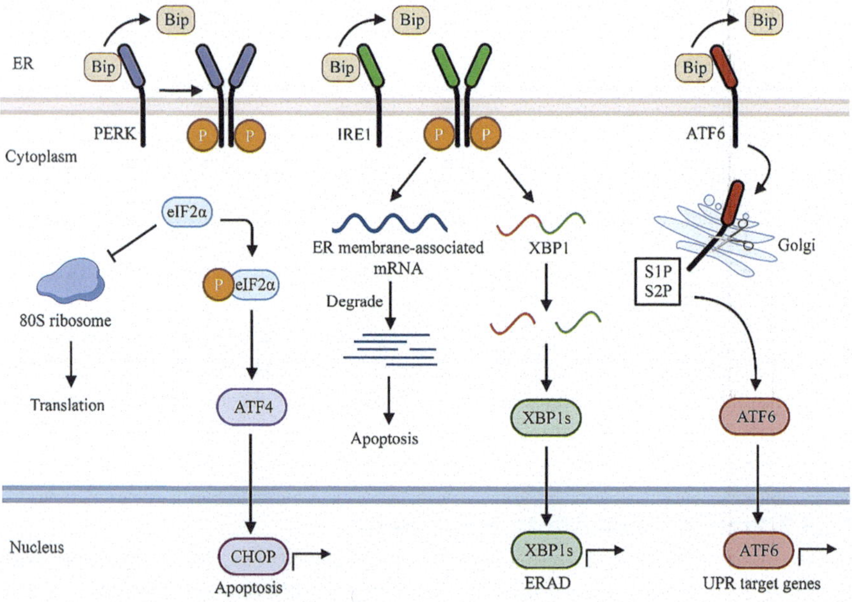

Fig. 5.17 UPR pathways. (Taken from Yuan et al. 2024 under CC-BY-4.0)

highlighted in the sections below reflecting the current interests of many biopharma and academic researchers.

The Unfolded Protein Response (UPR)

Improper folding of transmembrane and secreted proteins leads to their accumulation in the endoplasmic reticulum (ER) and contributes to diseases such as cancer, diabetes, and neurodegenerative disorders. The *unfolded protein response* (UPR) is an adaptive pathway employed by cells to restore correct protein misfolding or if this fails, to activate cell death mechanisms. The three main components of the adaptive pathway are shown in Fig. 5.17 (taken from Yuan et al. 2024).

PERK (protein kinase R-like endoplasmic reticulum kinase), IRE1 (inositol-requiring enzyme 1) and ATF6 (activating transcription factor 6) are bound to the *chaperone protein* Bip (immunoglobulin binding protein) which is released upon ER stress. PERK dimerises and phosphorylates eIF2α impeding the assembly of 80 S ribosomes and thereby inhibiting protein translation. Additionally, eIF2α induces the translation of activating transcription factor 4 (ATF4) and activates enhancer-binding protein C/EBP homologous protein (CHOP) expression, culminating in the induction of programmed cell death (*apoptosis*). Activated IRE1 dimerises and triggers its endonuclease activity thereby degrading mRNA in the ER and reducing protein biosynthesis. In addition, IRE1 cleaves the transcript encoding X box-binding protein 1 (XBP1), inducing endoplasmic reticulum-associated degradation (ERAD) gene expression, accelerating protein folding, and facilitating the degradation of misfolded proteins. ATF6 translocates to the Golgi, where it undergoes cleavage and activation by the S1 and S2 proteases, thereby activating the transcription of UPR target genes. While the UPR is important for maintaining cellular homeostasis, its over- activation contributes to disease through buildup of protein aggregates, inflammation, cell death, or dysregulation of cell growth and tumour formation.

The Heat Shock Response (HSR)

Cytosolic proteins are monitored by the *heat shock response* (HSR) in which cellular stress stimulates the production of heat shock factors (HSFs) such as HSP40, HSP60, HSP70, HSP90, and large HSPs that act as molecular chaperones to assist in correct protein folding (Zuo et al. 2024). The proteotoxic stress response is governed by heat shock transcription factors (HSFs) the master regulator being HSF-1 which produces multiple downstream effects on the HSR and other cellular processes (Chin et al. 2023).

Both the UPR and HSR mechanisms contain potential drug targets, particularly for oncology. Some small molecule development has been undertaken, for example for Bip, PERK and IRE1 inhibitors (Yuan et al. 2024) as well as HSP and HSF inhibitors (Zuo et al. 2024; Chin et al. 2023).

Given the importance of the unfolded protein response in normal cell function, it is not surprising that finding non-toxic inhibitors is challenging. However, proteostasis is implicated in resistance to the small molecule KRAS inhibitors discussed in Chap. 6. Drug resistance is an almost inevitable consequence of using signalling inhibitors of this type; inhibition of KRAS inactivates the heat shock and unfolded protein response via HSFG-1 and IRE1 resulting in cell death. However, resistant cells emerge by reactivating IRE1 through an ER stress-independent phosphorylation mechanism, essentially rewiring the pathway to stabilise the protein (Xiangdong et al. 2023). Targeting IRE1 could therefore be an effective strategy for overcoming resistance to KRAS inhibitors, supporting the potential of the unfolded protein response to offer novel drug targets in the future.

Subcellular Protein Localisation

Protein translation is confined to the cytosol but approximately half of the proteins generated by a cell have to be transported into or across at least one cellular membrane to reach their functional destination. Protein mislocalisation arises through a variety of mechanisms such as signal sequence mutations, changes in post-translational modifications or deregulation of the protein trafficking machinery (Hung and Link 2011).

A more recent publication by Ng et al. (2024) describes small molecule approaches for relocating misplaced proteins, essentially *treating disease through interactome rewiring*. Targeted relocalisation activating molecules (TRAMS) are bifunctional molecules that link the target protein to be relocated with a shuttle protein. Examples of the latter include nuclear hormone receptors that traffic between the cytoplasm and nucleus. The concept of designing small molecules to bring proteins into proximity is not new (see next section), but the approach of Ng et al. (2024) is a novel way of relocating proteins to their correct subcellular environment.

Targeted Protein Degradation (TPD) and Stabilisation (TPS)

The quality control of cellular proteins involves a highly regulated process of *waste disposal* based on the *ubiquitin-proteosome* system. Exploiting this system is one of the hottest areas in biopharmaceutical research since targeted degradation of a drug target protein (or its stabilisation) offers a powerful alternative to ligand binding or genetic regulation through gene therapy or RNA-based drugs. An overview of the ubiquitin-proteosome system taken from Kandel et al. (2024) is shown in Fig. 5.18.

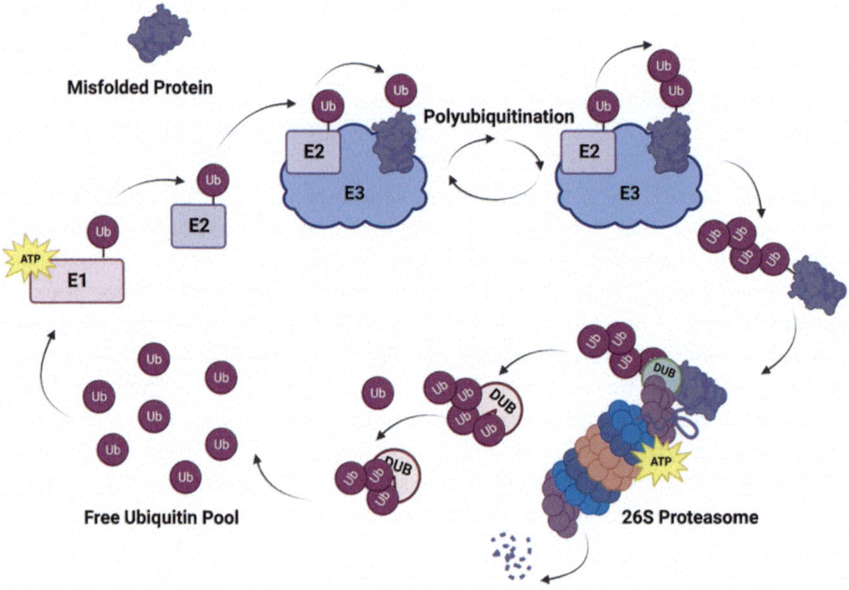

Fig. 5.18 The ubiquitin-proteosome system. (From Kandel et al. 2024 under CC-BY-4.0)

Ubiquitin (Ub) is a small 76 amino acid polypeptide covalently attached to a target protein through an isopeptide bond. The ubiquitination[22] acts as part of a signalling system with a number of outcomes, including targeted degradation as discussed here. Ub is initially conjugated to an E1 ubiquitin activating enzyme, in an ATP-dependent reaction, then transferred to an E2 Ub conjugation enzyme. E3 ligases bring an Ub-E2 complex in proximity with a target protein where Ub is transferred as a polyubiquitin chain to direct the modified protein to the multiprotein 20S proteasome complex. Folded proteins are processed by a 19S regulatory complex that delivers them to the 20S proteasome in an unfolded state. Deubiquitinases (DUBs) remove Ub before ATP- dependent proteasomal degradation. Other DUBs specifically degrade Ub chains to replenish the cellular pool of free ubiquitin, which is then accessible to E1 enzymes.

Protein modifications similar to ubiquitination include SUMOylation and NEDDylation, both of which may be important therapeutic targets in cancer. SUMO (small ubiquitin-like modifier) modifications can be disregulated in tumours with the relevant enzymes being potential small molecule drug targets (Kukkula et al. 2021).

NEDD8 (neural precursor cell expressed developmentally downregulated protein 8) is a ubiquitin-like protein that is covalently attached to a lysine residue of target proteins including cullins. In the latter case, neddylation leads to activation of the cullin-RING E3 ligases responsible for about 20% of cellular protein ubiquitination (Zhang et al. 2024). As with SUMOylation, inhibitors of enzymes involved with NEDDylation are being evaluated as anti-cancer drugs.

The human genome encodes two E1 enzymes, about 40 E2 enzymes, and approximately 600 E3 ligases, the latter allowing the ubiquitination of specific target proteins. This diversity of E3 ligases and their specificity for individual proteins has been exploited through the development of *Proteolysis targeting chimeras* (PROTACS) and *molecular glues* (MGs). Each brings drug target proteins in proximity to an E3 ligase thereby targeting the former for degradation by the proteosome (Zhong et al. 2024).

PROTACs are small molecules with a linker containing ligands for both the E3 ligase and target protein, while molecular glues are small molecules that serendipitously bind E3 and the target without an engineered linker. A prominent example of the latter is the drug thalidomide which gained notoriety through its teratogenic effects in pregnant women. Efforts to uncover its mechanism of action revealed it to bind to the E3 ligase cereblon (CRBN). Fortuitously thalidomide was effective in treating haematological malignancies like multiple myeloma through its targeted removal of IKAROS family zinc finger proteins prompting the clinical development of a derivative, lenalidomide.

The number of E3 ligases used by PROTACS and MGs is limited to CRBN, VHL, IAP MDM2 and DCAF15 which limits the possibilities for target degradation. This has led to efforts to find more targets for molecular glues within the human

[22] Sometimes referred to as ubiquitylation

genome, part of what Poirson et al. (2024) describe as *induced proximity proteomics*. Their unbiassed survey of the human proteome for potential molecular glue targets revealed E3 ligases as expected, but also surprisingly UBE2B, an E2 enzyme, plus other proteins unrelated to the ubiquitin-proteosome pathway. This opens up the possibility of making the search for molecular glues less of a random process and more of a systematic approach, as applied by PhoreMost in Cambridge, UK for example.

The opposite side of the coin to targeted protein degradation is targeted protein stabilisation. Deubiquitylating enzymes (DUBS) remove ubiquitin as part of the recycling process (see Fig. 5.18) so selectively activating this process could prevent degradation of a specific protein target. The human proteome contains around 100 DUBs representing six families based on sequence and domain conservation (Harrigan et al. 2018). Aberrant DUB activity is associated with a number of diseases, making these enzymes targets for small molecule discovery. By analogy with PROTACS and E3 ligases, deubiquitinase (DUB)-targeting chimeras (DUBTACS) are being developed to selectively stabilise protein targets (see commentary by Teh et al. 2022). This approach is showing some promise with cystic fibrosis in which stabilising the mutant ΔF508-CFTR protein led to improved function of the chloride channel altered in the disease. In the same proteome-wide approach for finding protein degraders, Poirson et al. (2024) have identified a number of stabilisers based on DUBs.

The repertoire of possible drug targets is being broadened from surface proteins alone to those inside the cell and those targetable with antibody constructs. Given the promise of targeted protein degradation and stabilisation in drug discovery, there is a great deal of academic and commercial activity in these areas a few examples of which are shown below:

The Centre for Protein Degradation ICR, London (2025)
CeTPD University of Dundee, Scotland https://www.dundee.ac.uk/cetpd Accessed 20 Mar 2025
Amphista Therapeutics Cambridge, UK (2025)
PhoreMost Cambridge, UK (2025)
The Center for Protein Degradation at Dana-Farber Cancer Institute, Boston, USA (2025)

Autophagy

Literally *eating oneself*, autophagy is cellular waste management system that maintains cell viability, unlike the cell death pathways discussed later.

Macroautophagy/autophagy is an evolutionarily conserved pathway responsible for clearing cytosolic aggregated proteins, damaged organelles or invading microorganisms. Dysfunctional autophagy leads to pathological accumulation of the cargo which has been linked to a range of human diseases, including neurodegenerative diseases, infectious and autoimmune diseases and various forms of cancer (Kocak et al. 2021). In macroautophagy, targets to be degraded are sequestered in autophagosomes that fuse with lysosomes or late endosomes to trap the target molecules prior to degrading them with a variety of lysosomal acid hydrolases.

The signalling molecules involved in these processes have been described in detail, for example the activating pathway PI3K-AKT-MTOR that can be targeted by rapamycin. However, specific autophagy targets have not currently been developed, which creates opportunities for future research. One way forward may be through targeting autophagy proteins with protein degraders similar to PROTACs. Kocak et al. (2021) have described a series of such molecules with targeted degradation in mind. In addition to PROTACs these are categorised as LYTACs (lysosome targeting chimeras), AUTACs (autophagy-targeting chimeras) and ATTECs (autophagosome-tethering compounds).

5.3.4.3 Cell Death Pathways

Cell death is an intrinsic aspect of organismal homeostasis and can either be the result of a programmed mechanism, or from inflammation due to tissue injury. Not surprisingly, cell death pathways are a source of drug targets, both in encouraging cell death (e.g. in cancer) or protecting cells during immunoinflammatory responses. A seemingly unstoppable number of different cell death pathways are described in the literature, each with individual biochemical and subcellular changes. One of the most recent is *triaptosis* involves PI-3 kinase in responses to antioxidants (Swamynathan et al. 2024). Park et al. (2023) have reviewed the other pathways in some detail (listed in Table 5.12).

Table 5.12 Different cell death pathways

	Pathway	Description
RCD	Apoptosis (Anoikis)	Organised cell breakdown
" "	Autophagic cell death (autosis)	Autophagosome accumulation
" "	Entosis	Engulfment of cells resulting in their destruction
" "	Methuosis	Swelling of macropinosomal vacuoles
" "	Paraptosis	Swelling of vacuoles
" "	Mitoptosis	Mitochondrial suicide
" "	Parthanatos	PARP activation
" "	Ferroptosis	Iron-dependent oxidative stress
" "	Cuproptosis	Copper induced mitochondrial damage
" "	Necroptosis	Necrosome formation
" "	Lysosomal cell death	Lysosomal membrane permeabilization
" "	Pyroptosis	Activation of immune cell death
" "	NETosis	Release of neutrophil extracellular traps (NETs)
Non-RCD	Necrosis	Release of cell contents
Other	Alkaliptosis	Alkali-induced necrosis
" "	Oxeiptosis	Noninflammatory cell death pathway
" "	Erebosis	Gut enterocyte cell death with no distinguishing features

From: Park et al. (2023) under CC-BY-4.0. *RCD* regulated cell death

Apoptosis

We are most grateful to Professor James Cormack of the Department of Greek, University of Aberdeen, for suggesting this term. The word "apoptosis"(ἀπόπτωσις) is used in Greek to describe the "dropping off" or "falling off" of petals from flowers, or leaves from trees.

So concludes the paper by Kerr et al. (1972) describing the characteristics of the programmed cell death mechanism used to, amongst other things, regulate blood cell numbers, sculpt tissues during embryogenesis and respond to cancer chemotherapy. Apoptosis can be initiated by both extrinsic signals (death receptors pathway) and intrinsic signals (mitochondrial responses) which activate the caspase family of proteases. Apoptosis has been of interest to cancer researchers for years since tumour cells often express high levels of anti-apoptotic proteins and become *addicted* to them for their survival. Consequently, anti-apoptotic proteins (predominantly IAPs and Bcl-2 family members) have emerged as attractive druggable targets for the development of cancer therapeutics (Shahar and Larisch 2020). This extends to inducing apoptosis in *senescent cells* that have reached the limits of cell division which are not removed from tissues but remain to cause damage through cytokine release etc. Targeting senescent cells by inhibiting the antiapoptotic protein Bcl-xL, for example could slow down the physical degeneration associated with ageing, which of course is a major area of biopharmaceutical interest (Wong 2024).

Although candidate inhibitors of the above proteins have been deployed in the clinic, there is room for novel approaches including manipulating protein proximity as discussed in Sect. 5.3.4.2. For example, Sarotte et al. (2024) have created bifunctional small molecules that link transcriptional kinases to the anti-apoptotic protein BCL6 thereby removing co-repressors from the latter and promoting the activation of proapoptotic genes. Other scenarios could be envisaged such as using targeted protein degradation strategies to remove BCL and IAP family members.

Non-apoptotic Cell Death

Apoptosis and necrosis are major cell death pathways, but looking at Table 5.12, clearly not the only ones; there are plenty of opportunities for target discovery, particularly for cancer, where resistance to apoptosis is a problem (see above). However, targeting multiple cell death pathways using combination treatments could overcome this obstacle (Hadian and Stockwell 2023).

5.3.4.4 Cell Biology and Machine Learning

Cell biology is fundamentally limited in its ability to collect complete data on cellular phenotypes and the wide range of responses to perturbation. Areas such as computer vision and speech recognition have addressed this problem of characterizing unseen or unlabelled conditions with the combined advances of big data, deep learning, and computing resources in the past 5 years. Similarly, recent advances in machine learning approaches enabled by single-cell data have begun to address prediction tasks in perturbation response modelling (Ji et al. 2021).

In other words, a major objective in drug discovery is the ability to predict outcomes of a therapeutic intervention, both in terms of disease modification and adverse effects. Despite the major advances described earlier, it will never be possible to measure all the variables described by genomics, proteomics and metabolomics etc., but computational tools based on deep learning should begin to fill in the gaps as is the case in other fields of machine learning. Uhler (2024) has highlighted the advances in machine learning that will be required to infer causal mechanisms of disease as opposed to correlations where there is no need to understand how the result was achieved.

Inferring causal relationships using machine learning is difficult, but with new computational tools and large experimental datasets, it is not impossible. One challenge is to extract meaning from datasets generated on from completely different experimental platforms, single cell sequencing and histological images, for example. In the latter case a future scenario might occur in which an investigator producing a haematoxylin and eosin (H&E) tissue image can ask the computer to explain which pathways or gene modules are perturbed (Uhler 2024).

Research into solving these problems is ongoing (e.g. Yang et al. 2021) and there is every reason to believe that target discovery in the near and foreseeable future will occur not only through serendipity but also through systematic investigation of biological and disease pathways at levels of organisation ranging from the whole organism to individual molecules.

References

Abramson J et al (2024) Accurate structure prediction of biomolecular interactions with AlphaFold 3. Nature 630:493–500. https://doi.org/10.1038/s41586-024-07487-w. Accessed 20 Mar 2025

Aldridge S, Teichmann SA (2020) Single cell transcriptomics comes of age. Nat Commun 11:4307. https://doi.org/10.1038/s41467-020-18158-5. Accessed 20 Mar 2025

Alexandrov LB et al (2020) The repertoire of mutational signatures in human cancer. Nature 578: 94–101. https://doi.org/10.1038/s41586-020-1943-3. Accessed 20 Mar 2025

AlphaFold Protein Structure Database (2025). https://alphafold.ebi.ac.uk/. Accessed 20 Mar 2025

Amphista Therapeutics Cambridge, UK (2025). https://amphista.com/. Accessed 20 Mar 2025

Ariceum (n.d.). https://ariceum-therapeutics.com/ariceum-therapeutics-commences-phase-1-clinical-trial-in-patients-with-recurrent-glioblastoma/. Accessed 20 Mar 2025

Arima (2025). https://arimagenomics.com/applications/genome-assembly/. Accessed 20 Mar 2025

Avram S et al (2024) Novel drug targets in 2023. Nat Rev Drug Dis 23:330. https://doi.org/10.1038/d41573-024-00057-9. Accessed 20 Mar 2025

Azinas S, Carroni M (2023) Cryo-EM uniqueness in structure determination of macromolecular complexes: a selected structural anthology. Curr Opin Struct Biol 81:102621. https://doi.org/10.1016/j.sbi.2023.102621. Accessed 20 Mar 2025

Baek M et al (2021) Accurate prediction of protein structures and interactions using a three-track neural network. Science 373:871–876. https://doi.org/10.1126/science.abj8754. Accessed 20 Mar 2025

Balasubramanian H et al (2023) Imagining the future of optical microscopy: everything, everywhere, all at once. Commun Biol 6:1096. https://doi.org/10.1038/s42003-023-05468-9. Accessed 20 Mar 2025

Baldwin ET et al (2024) Structures, functions and adaptations of the human LINE-1 ORF2 protein. Nature 626:194–206. https://doi.org/10.1038/s41586-023-06947-z. Accessed 20 Mar 2025

Bamshad MJ et al (2019) Mendelian gene discovery: fast and furious with no end in sight. Am J Hum Genet 105:448–455. https://doi.org/10.1016/j.ajhg.2019.07.011. Accessed 20 Mar 2025

Ban TA (2022) The role of serendipity in drug discovery. Dialogues Clin Neurosci 8:335–344. https://doi.org/10.31887/DCNS.2006.8.3/tban. Accessed 20 Mar 2025

Batista P, Chang HY (2013) Long noncoding RNAs: cellular address codes in development and disease. Cell 152:1298–1307. https://doi.org/10.1016/j.cell.2013.02.012. Accessed 20 Mar 2025

Bravo JPK et al (2024) Plasmid targeting and destruction by the DdmDE bacterial defence system. Nature 630:961–967. https://doi.org/10.1038/s41586-024-07515-9. Accessed 20 Mar 2025

Brezgin S et al (2019) Dead cas systems: types, principles, and applications. Int J Mol Sci 20:6041. https://doi.org/10.3390/ijms20236041. Accessed 20 Mar 2025

Callaway E (2024a) AlphaFold found thousands of possible psychedelics. Will its predictions help drug discovery? Nature 626:14–15. https://doi.org/10.1038/d41586-024-00130-8. Accessed 20 Mar 2025

Callaway E (2024b) 'ChatGPT for CRISPR' creates new gene-editing tools. Nature 629:272. https://doi.org/10.1038/d41586-024-01243-w. Accessed 20 Mar 2025

Cancer Grand Challenges – Mutagraphs (2024). https://www.cancergrandchallenges.org/mutographs. Accessed 9 Oct 2024

Cancer Grand Challenges- Normal Phenotypes (2025). https://www.cancergrandchallenges.org/challenges/active-challenges/normal-phenotypes. Accessed 20 Mar 2025

Carpenter AE, Cimini BA, Eliceiri KW (2023) Smart microscopes of the future. Nat Methods 20(7): 962–964. https://doi.org/10.1038/s41592-023-01912-0. Accessed 20 Mar 2025

Chan EM et al (2019) WRN helicase is a synthetic lethal target in microsatellite unstable cancers. Nature 568:551–556. https://doi.org/10.1038/s41586-019-1102-x. Accessed 20 Mar 2025

Chen B et al (2018) Top-down proteomics: ready for prime time? Anal Chem 90:110–127. https://doi.org/10.1021/acs.analchem.7b04747. Accessed 20 Mar 2025

Chin Y et al (2023) Targeting HSF1 for cancer treatment: mechanisms and inhibitor development. Theranostics 13:2281–2300. https://doi.org/10.7150/thno.82431. Accessed 20 Mar 2025

Clark IC et al (2023) Microfluidics-free single-cell genomics with templated emulsification. Nat Biotechnol 41:1557–1566. https://doi.org/10.1038/s41587-023-01685-z. Accessed 20 Mar 2025

Claussnitzer M et al (2020) A brief history of human disease genetics. Nature 577:179–189. https://doi.org/10.1038/s41586-019-1879-7. Accessed 20 Mar 2025

Cornish AJ et al (2024) The genomic landscape of 2,023 colorectal cancers. Nature 633:127–136. https://doi.org/10.1038/s41586-024-07747-9. Accessed 9 Oct 2024

COSMIC Mutational Signatures, Sanger Centre (2025). https://cancer.sanger.ac.uk/signatures/. Accessed 20 Mar 2025

Costanzi S et al (2009) Rhodopsin and the others: a historical perspective on structural studies of G protein-coupled receptors. Curr Pharm Des 15:3994–4002. https://doi.org/10.2174/138161209789824795. Accessed 20 Mar 2025

Cunningham M et al (2023) PINNED: identifying characteristics of druggable human proteins using an interpretable neural network. J Cheminform 15:64. https://doi.org/10.1186/s13321-023-00735-7. Accessed 20 Mar 2025

Curry S (2015) Structural biology: a century-long journey into an unseen world. Interdiscip Sci Rev 40:308–328. https://doi.org/10.1179/0308018815Z.000000000120. Accessed 20 Mar 2025

CZ CELLxGENE Discover (2025). https://cellxgene.cziscience.com/. Accessed 20 Mar 2025

Dabbagh-Gorjani F (2024) A comprehensive review on the role of Interleukin-40 as a biomarker for diagnosing inflammatory diseases. Autoimmune Dis. https://doi.org/10.1155/2024/3968767

Dance A (2024) Computational technologies of the human cell atlas. Nature 635:773–775. https://doi.org/10.1038/d41586-024-03762-y. Accessed 20 Mar 2025

Della Badia LA et al (2016) Targeting PCSK9 as a promising new mechanism for lowering low-density lipoprotein cholesterol. Pharmacol Ther 164:183–194. https://doi.org/10.1016/j.pharmthera.2016.04.011. Accessed 20 Mar 2025

Doron S et al (2018) Systematic discovery of antiphage defense systems in the microbial pangenome. Science 359:eaar4120. https://doi.org/10.1126/science.aar4120. Accessed 20 Mar 2025

Drews J, Ryser S (1997) The role of innovation in drug development. Nat Biotechnol 15:1318–1319. https://doi.org/10.1038/nbt1297-1318. Accessed 20 Mar 2025

Dupont C et al (2009) Epigenetics: definition, mechanisms and clinical perspective. Semin Reprod Med 27:351–357. https://doi.org/10.1055/s-0029-1237423. Accessed 20 Mar 2025

Durrant MG et al (2024) Bridge RNAs direct programmable recombination of target and donor DNA. Nature 630:984–993. https://doi.org/10.1038/s41586-024-07552-4. Accessed 20 Mar 2025

Eisenstein M (2022a) Adrestia therapeutics – gene networks to the rescue. Nat Biotech. https://doi.org/10.1038/d41587-022-00011-3

Eisenstein M (2022b) Machine learning powers biobank-driven drug discovery. Nat Biotech 40: 1303–1305. https://doi.org/10.1038/s41587-022-01457-1. Accessed 20 Mar 2025

Eisenstein M (2024) Startups probe hidden viruses in the 'dark genome' to treat disease. Nat Biotechnol 42:539–545. https://doi.org/10.1038/s41587-024-02215-1. Accessed 20 Mar 2025

Enhanc3D Genomics (2025). https://enhanc3dgenomics.com/. Accessed 20 Mar 2025

eQTLGen Consortium (2025). https://eqtlgen.org/phase1.html. Accessed 20 Mar 2025

Ferkingstad E et al (2021) Large-scale integration of the plasma proteome with genetics and disease. Nat Genet 53:1712–1721. https://doi.org/10.1038/s41588-021-00978-w. Accessed 20 Mar 2025

Ferretti S et al (2024) Discovery of WRN inhibitor HRO761 with synthetic lethality in MSI cancers. Nature 629:443–449. https://doi.org/10.1038/s41586-024-07350-y. Accessed 20 Mar 2025

Finan C et al (2017) The druggable genome and support for target identification and validation in drug development. Sci Transl Med 9:383. https://doi.org/10.1126/scitranslmed.aag1166. Accessed 20 Mar 2025

Finngen (2025). https://www.finngen.fi/en. Accessed 20 Mar 2025

Fu L et al (2023) Fibrinogen-like protein 2 in inflammatory diseases: a future therapeutic target. Int Immunopharmacol 116:109799. https://doi.org/10.1016/j.intimp.2023.109799. Accessed 20 Mar 2025

Garana BB, Graham NA (2022) Metabolomics paves the way for improved drug target identification. Mol Syst Biol 18:e10914. https://doi.org/10.15252/msb.202210914. Accessed 20 Mar 2025

Gemmer M et al (2023) Visualization of translation and protein biogenesis at the ER membrane. Nature 614:160–167. https://doi.org/10.1038/s41586-022-05638-5. s

gnomAD (2025). https://gnomad.broadinstitute.org/about. Accessed 20 Mar 2025

Greber BJ, Nogales E (2019) The structures of eukaryotic transcription pre-initiation complexes and their functional implications. In: Harris J, Marles-Wright J (eds) Macromolecular protein complexes II: structure and function. Subcellular biochemistry, vol 93. Springer, Cham. https://doi.org/10.1007/978-3-030-28151-9_5. Accessed 20 Mar 2025

GTEx Project (2025). https://www.gtexportal.org/home/aboutAdultGtex. Accessed 20 Mar 2025

Hadian K, Stockwell BR (2023) The therapeutic potential of targeting regulated non-apoptotic cell death. Nat Rev Drug Discov 22:723–742. https://doi.org/10.1038/s41573-023-00749-8. Accessed 20 Mar 2025

Halldorsson BV et al (2022) The sequences of 150,119 genomes in the UK biobank. Nature 607: 732–740. https://doi.org/10.1038/s41586-022-04965-x. Accessed 20 Mar 2025

Halliwell RF (2007) A short history of the rise of the molecular pharmacology of ionotropic drug receptors. Trends Pharmacol Sci 28:214–219. https://doi.org/10.1016/j.tips.2007.03.007. Accessed 20 Mar 2025

Hampton HG et al (2020) The arms race between bacteria and their phage foes. Nature 577:327–336. https://doi.org/10.1038/s41586-019-1894-8. Accessed 20 Mar 2025

Harrigan J et al (2018) Deubiquitylating enzymes and drug discovery: emerging opportunities. Nat Rev Drug Discov 17:57–78. https://doi.org/10.1038/nrd.2017.152. Accessed 20 Mar 2025

Harvard Chan Bioinformatics Core (2025). https://bioinformatics.sph.harvard.edu/. Accessed 20 Mar 2025
He Y et al (2024a) A comprehensive whole genome database of ethnic minority populations. Sci Rep 14:13954. https://doi.org/10.1038/s41598-024-63892-1. Accessed 20 Mar 2025
He F et al (2024b) Π-HuB: the proteomic navigator of the human body. Nature 636:322–331. https://doi.org/10.1038/s41586-024-08280-5. Accessed 20 Mar 2025
HGNC (2024). https://www.genenames.org/download/statistics-and-files/. Accessed 20 Mar 2025
Höhn A, Tramutola A, Cascella R (2020) Proteostasis failure in neurodegenerative diseases: focus on oxidative stress. Oxidative Med Cell Longev 2020(1):5497046. https://doi.org/10.1155/2020/5497046. Accessed 20 Mar 2025
Hu T et al (2023) Single-cell spatial metabolomics with cell-type specific protein profiling for tissue systems biology. Nat Commun 14:8260. https://doi.org/10.1038/s41467-023-43917-5. Accessed 20 Mar 2025
Hukerikar N et al (2024) Prioritising genetic findings for drug target identification and validation. Atherosclerosis 390:117462. https://doi.org/10.1016/j.atherosclerosis.2024.117462. Accessed 20 Mar 2025
Human Cell Atlas (2025). https://www.humancellatlas.org/. Accessed 20 Mar 2025
Hung M-C, Link W (2011) Protein localization in disease and therapy. J Cell Sci 124:3381–3392. https://doi.org/10.1242/jcs.089110. Accessed 20 Mar 2025
Illumina (2025). https://www.illumina.com/. Accessed 20 Mar 2025
Illumina Techniques (2025). https://emea.illumina.com/techniques/popular-applications/qtl-analysis.html. Accessed 20 Mar 2025
Jackson S, Bartek J (2009) The DNA-damage response in human biology and disease. Nature 461:1071–1078. https://doi.org/10.1038/nature08467. Accessed 20 Mar 2025
Javierre BM et al (2016) Lineage-specific genome architecture links enhancers and non-coding disease variants to target gene promoters. Cell 167:1369–1384. https://doi.org/10.1016/j.cell.2016.09.037. Accessed 20 Mar 2025
Jerby-Arnon L et al (2014) Predicting cancer-specific vulnerability via data-driven detection of synthetic lethality. Cell 158:1199–1209. https://doi.org/10.1016/j.cell.2014.07.027. Accessed 20 Mar 2025
Ji Y et al (2021) Machine learning for perturbational single-cell omics. Cell Systems 12:522–537. https://doi.org/10.1016/j.cels.2021.05.016. Accessed 20 Mar 2025
Jumper J et al (2021) Highly accurate protein structure prediction with AlphaFold. Nature 596:583–589. https://doi.org/10.1038/s41586-021-03819-2. Accessed 20 Mar 2025
Kalidas Y, Chandra N (2008) PocketDepth: a new depth based algorithm for identification of ligand binding sites in proteins. J Struct Biol 161:31–42. https://doi.org/10.1016/j.jsb.2007.09.005. Accessed 20 Mar 2025
Kandel R et al (2024) Proteotoxic stress and the ubiquitin proteasome system. Semin Cell Dev Biol 156:107–120. https://doi.org/10.1016/j.semcdb.2023.08.002. Accessed 20 Mar 2025
Karras P et al (2024) Decoding the interplay between genetic and non-genetic drivers of metastasis. Nature 629:543–554. https://doi.org/10.1038/s41586-024-07302-6. Accessed 20 Mar 2025
Kerr J et al (1972) Apoptosis: a basic biological phenomenon with Wideranging implications in tissue kinetics. Br J Cancer 26:239–257. https://doi.org/10.1038/bjc.1972.33. Accessed 20 Mar 2025
Kim MS et al (2014) A draft map of the human proteome. Nature 509:575–581. https://doi.org/10.1038/nature13302. Accessed 20 Mar 2025
King EA et al (2019) Are drug targets with genetic support twice as likely to be approved? Revised estimates of the impact of genetic support for drug mechanisms on the probability of drug approval. PLoS Genet 15:e1008489. https://doi.org/10.1371/journal.pgen.1008489. Accessed 20 Mar 2025
Kjer-Hansen P et al (2024) Protein isoform-centric therapeutics: expanding targets and increasing specificity. Nat Rev Drug Discov 23:759–779. https://doi.org/10.1038/s41573-024-01025-z. Accessed 20 Mar 2025

Kocak M et al (2021) Targeting autophagy in disease: established and new strategies. Autophagy 18:473–495. https://doi.org/10.1080/15548627.2021.1936359. Accessed 20 Mar 2025
Krishna R et al (2024) Generalized biomolecular modeling and design with RoseTTAFold all-atom. Science 384. https://doi.org/10.1126/science.adl2528. Accessed 20 Mar 2025
Krumm A, Duan Z (2019) Understanding the 3D genome: emerging impacts on human disease. Semin Cell Dev Biol 90:62–77. https://doi.org/10.1016/j.semcdb.2018.07.004. Accessed 20 Mar 2025
Kukkula A et al (2021) Therapeutic potential of targeting the SUMO pathway in cancer. Cancers 13:4402. https://doi.org/10.3390/cancers13174402. Accessed 20 Mar 2025
Lander ES (2016) The heroes of CRISPR. Cell 164:18–28. https://doi.org/10.1016/j.cell.2015.12.041. Accessed 20 Mar 2025
Lau AM et al (2024) Exploring structural diversity across the protein universe with the encyclopedia of domains. Science 386. https://doi.org/10.1126/science.adq4946. Accessed 20 Mar 2025
Ledford H (2024) No CRISPR: oddball 'jumping gene' enzyme edits genomes without breaking DNA. Nature 631:470–471. https://doi.org/10.1038/d41586-024-02141-x. Accessed 20 Mar 2025
Li Y-C et al (2004) Microsatellites within genes: structure, function, and evolution. Mol Biol Evol 21:991–1007. https://doi.org/10.1093/molbev/msh073. Accessed 20 Mar 2025
Lindon JC et al (2003) So what's the deal with metabonomics? Anal Chem 75. https://doi.org/10.1021/ac031386+. Accessed 20 Mar 2025
Liu Q et al (2022) Alternative splicing and isoforms: from mechanisms to diseases. Genes 13:401. https://doi.org/10.3390/genes13030401. Accessed 20 Mar 2025
Liu F et al (2023) Technologies and applications of single-cell DNA methylation sequencing. Theranostics 13:2439–2454. https://doi.org/10.7150/thno.82582. Accessed 20 Mar 2025
Lonsdale J et al (2013) The genotype-tissue expression (GTEx) project. Nat Genet 45:580–585. https://doi.org/10.1038/ng.2653. Accessed 20 Mar 2025
Lopez SC et al (2022) Precise genome editing across kingdoms of life using retron-derived DNA. Nat Chem Biol 18:199–206. https://doi.org/10.1038/s41589-021-00927-y. Accessed 20 Mar 2025
LoQus23 (2025). https://loqus23.com/. Accessed 20 Mar 2025
Macosko EZ et al (2015) Highly parallel genome-wide expression profiling of individual cells using nanoliter droplets. Cell 161:1202–1214. https://doi.org/10.1016/j.cell.2015.05.002. Accessed 20 Mar 2025
Mahdessian D et al (2021) Spatiotemporal dissection of the cell cycle with single-cell proteogenomics. Nature 590:649–654. https://doi.org/10.1038/s41586-021-03232-9. Accessed 20 Mar 2025
Makarova KS et al (2020) Evolutionary classification of CRISPR–Cas systems: a burst of class 2 and derived variants. Nat Rev Microbiol 18:67–83. https://doi.org/10.1038/s41579-019-0299-x. Accessed 20 Mar 2025
Martínez-Jiménez F et al (2023) Pan-cancer whole-genome comparison of primary and metastatic solid tumours. Nature 618:333–341. https://doi.org/10.1038/s41586-023-06054-z. Accessed 20 Mar 2025
Masanori I et al (2005) Genome-wide application of RNAi to the discovery of potential drug targets. FEBS Lett 579:5988–5995. https://doi.org/10.1016/j.febslet.2005.08.015. Accessed 20 Mar 2025
McIlwain S et al (2020) Enhancing top-down proteomics data analysis by combining deconvolution results through a machine learning strategy. J Am Soc Mass Spectrom 31:1104–1113. https://doi.org/10.1021/jasms.0c00035. Accessed 20 Mar 2025
Mertz L (2023) Advances in Microscopy Tech Offer Better Views. IEEE Pulse. https://www.embs.org/pulse/articles/advances-in-microscopy-tech-offer-better-views/. Accessed 20 Mar 2025
Michoel T, Zhang JD (2023) Causal inference in drug discovery and development. Drug Discov Today 28:103737. https://doi.org/10.1016/j.drudis.2023.103737. Accessed 20 Mar 2025

Minikel VE et al (2024) Refining the impact of genetic evidence on clinical success. Nature 629: 624–629. https://doi.org/10.1038/s41586-024-07316-0. Accessed 20 Mar 2025

Morgan DO (2007) The Cell Cycle: Principles of Control. https://morganlab.ucsf.edu/cell-cycle-principles-control. Accessed 20 Mar 2025

Morrow JF et al (1974) Replication and transcription of eukaryotic DNA in Escherichia coli. Proc Natl Acad Sci 71:1743–1747. https://doi.org/10.1073/pnas.71.5.1743. Accessed 20 Mar 2025

Motone K et al (2024) Multi-pass, single-molecule nanopore reading of long protein strands. Nature 633:662–669. https://doi.org/10.1038/s41586-024-07935-7. Accessed 20 Mar 2025

Naito T, Okada Y (2024) Genotype imputation methods for whole and complex genomic regions utilizing deep learning technology. J Hum Genet 69:481–486. https://doi.org/10.1038/s10038-023-01213-6. Accessed 20 Mar 2025

Nemudrala A et al (2024) Repair of CRISPR-guided RNA breaks enables site-specific RNA excision in human cells. Science 384:808–814. https://doi.org/10.1126/science.adk5518. Accessed 20 Mar 2025

Ng CSC et al (2024) Targeted protein relocalization via protein transport coupling. Nature 633:941–951. https://doi.org/10.1038/s41586-024-07950-8. Accessed 20 Mar 2025

NHS Genomic Medicine Service (2025). https://www.england.nhs.uk/genomics/nhs-genomic-med-service/. Accessed 20 Mar 2025

Nogales E, Mahamid J (2024) Bridging structural and cell biology with cryo-electron microscopy. Nature 628:47–56. https://doi.org/10.1038/s41586-024-07198-2. Accessed 20 Mar 2025

Nurk S et al (2022) The complete sequence of a human genome. Science 376:44–53. https://doi.org/10.1126/science.abj6987. Accessed 20 Mar 2025

Obeidat M et al (2021) A comprehensive evaluation of potential lung function associated genes in the SpiroMeta general population sample. PLoS One 6:e19382. https://doi.org/10.1371/journal.pone.0019382. Accessed 20 Mar 2025

Oguchi A et al (2024) An atlas of transcribed enhancers across helper T cell diversity for decoding human diseases. Science 385:43. https://doi.org/10.1126/science.add8394. Accessed 20 Mar 2025

OMIM Online Mendelian Inheritance in Man (2025). https://www.omim.org/help/about. Accessed 20 Mar 2025

Osterman I, Sorek R (2024) Tricking phages with a reverse move. Science 386:25–26. https://doi.org/10.1126/science.ads3638. Accessed 20 Mar 2025

Oxford Nanopore (2025). https://nanoporetech.com/. Accessed 20 Mar 2025

PacBio (2025). https://www.pacb.com/products-and-services/applications/whole-genome-sequencing/. Accessed 20 Mar 2025

Palstra R-J, Grosveld F (2012) Transcription factor binding at enhancers: shaping a genomic regulatory landscape in flux. Front Genet 3:195. https://doi.org/10.3389/fgene.2012.00195. Accessed 20 Mar 2025

Pan S-Y et al (2013) New perspectives on how to discover drugs from herbal medicines: CAM′ s outstanding contribution to modern therapeutics. Evid Based Complement Alternat Med. https://doi.org/10.1155/2013/627375. Accessed 20 Mar 2025

Pan S et al (2024) Metabolomics-driven approaches for identifying therapeutic targets in drug discovery. Med Comm 5:e792. https://doi.org/10.1002/mco2.792. Accessed 20 Mar 2025

Park W et al (2023) Diversity and complexity of cell death: a historical review. Exp Mol Med 55: 1573–1594. https://doi.org/10.1038/s12276-023-01078-x. Accessed 20 Mar 2025

Pennisi E (2024) Dark proteome' survey reveals thousands of new human genes. Science 386:951–952. https://doi.org/10.1126/science.zxaor8l. Accessed 20 Mar 2025

Petropoulos M et al (2024) Transcription–replication conflicts underlie sensitivity to PARP inhibitors. Nature 628:433–441. https://doi.org/10.1038/s41586-024-07217-2. Accessed 20 Mar 2025

Petrosius V, Schoof EM (2023) Recent advances in the field of single-cell proteomics. Transl Oncol 27:101556. https://doi.org/10.1016/j.tranon.2022.101556. Accessed 20 Mar 2025

PhoreMost Cambridge, UK (2025). https://www.phoremost.com/science. Accessed 20 Mar 2025

Poirson J et al (2024) Proteome-scale discovery of protein degradation and stabilization effectors. Nature 628:878–886. https://doi.org/10.1038/s41586-024-07224-3. Accessed 20 Mar 2025
Previtali V et al (2024) New horizons of synthetic lethality in cancer: current development and future perspectives. J Med Chem 67:11488–11521. https://doi.org/10.1021/acs.jmedchem.4c00113. Accessed 20 Mar 2025
Proteostasis Consortium (2025). https://www.proteostasisconsortium.com. Accessed 20 Mar 2025
Prüll C-R et al (2009) A short history of the drug receptor concept. Palgrave Macmillan, Basingstoke. https://doi.org/10.1057/9780230583740
Qi T et al (2022) Genetic control of RNA splicing and its distinct role in complex trait variation. Nat Genet 54:1355–1363. https://doi.org/10.1038/s41588-022-01154-4. Accessed 20 Mar 2025
Rask-Andersen M et al (2011) Trends in the exploitation of novel drug targets. Nat Rev Drug Discov 10:579–590. https://doi.org/10.1038/nrd3478. Accessed 20 Mar 2025
RCSB PDB (2025). https://www.rcsb.org/. Accessed 20 Mar 2025
Rhie A et al (2023) The complete sequence of a human Y chromosome. Nature 621:344–354. https://doi.org/10.1038/s41586-023-06457-y. Accessed 20 Mar 2025
Risca VI, Greenleaf WJ (2015) Unraveling the 3D genome: genomics tools for multiscale exploration. Trends Genet 31:357–372. https://doi.org/10.1016/j.tig.2015.03.010. Accessed 20 Mar 2025
Ruffolo JA et al (2024) Design of highly functional genome editors by modeling the universe of CRISPR-Cas sequences. https://doi.org/10.1101/2024.04.22.590591. Accessed 20 Mar 2025
Santos R et al (2017) A comprehensive map of molecular drug targets. Nat Rev Drug Discov 16:19–34. https://doi.org/10.1038/nrd.2016.230. Accessed 20 Mar 2025
Sarotte RC et al (2024) Relocalizing transcriptional kinases to activate apoptosis. Science 386. https://doi.org/10.1126/science.adl5361. Accessed 20 Mar 2025
Schäffer A et al (2024) A systematic analysis of the landscape of synthetic lethality-driven precision oncology. Med 5:73–89. https://doi.org/10.1016/j.medj.2023.12.009. Accessed 20 Mar 2025
Schiesser S et al (2022) Shedding light on the dark genome: drugging long non-coding RNA. Future Drug Discov 4(2). https://doi.org/10.4155/fdd-2022-0009. Accessed 20 Mar 2025
Schmidt AF et al (2020) Genetic drug target validation using Mendelian randomisation. Nat Commun 11:3255. https://doi.org/10.1038/s41467-020-16969-0. Accessed 20 Mar 2025
Senkin S et al (2024) Geographic variation of mutagenic exposures in kidney cancer genomes. Nature 629:910–918. https://doi.org/10.1038/s41586-024-07368-2. Accessed 20 Mar 2025
Shahar N, Larisch S (2020) Inhibiting the inhibitors: targeting anti-apoptotic proteins in cancer and therapy resistance. Drug Resist Updat 52:100712. https://doi.org/10.1016/j.drup.2020.100712. Accessed 20 Mar 2025
Sinitcyn P et al (2023) Global detection of human variants and isoforms by deep proteome sequencing. Nat Biotechnol 41:1776–1786. https://doi.org/10.1038/s41587-023-01714-x. Accessed 20 Mar 2025
Spurious Correlations (2025). https://tylervigen.com/spurious-correlations. Accessed 20 Mar 2025
Storm Therapeutics (2025). https://www.stormtherapeutics.com/clinical-trials/. Accessed 20 Mar 2025
Sun KY et al (2024) A deep catalogue of protein-coding variation in 983,578 individuals. Nature 631:583–592. https://doi.org/10.1038/s41586-024-07556-0. Accessed 20 Mar 2025
Sundaram L et al (2024) Single-cell chromatin accessibility reveals malignant regulatory programs in primary human cancers. Science 385:1–16. https://doi.org/10.1126/science.adk9217. Accessed 20 Mar 2025
Swamynathan MM et al (2024) Dietary pro-oxidant therapy by a vitamin K precursor targets PI 3-kinase VPS34 function. Science 386. https://doi.org/10.1126/science.adk9167. Accessed 20 Mar 2025
Swarts D et al (2014) The evolutionary journey of Argonaute proteins. Nat Struct Mol Biol 21:743–753. https://doi.org/10.1038/nsmb.2879. Accessed 20 Mar 2025
Taylor DJ et al (2024) Sources of gene expression variation in a globally diverse human cohort. Nature 632:122–130. https://doi.org/10.1038/s41586-024-07708-2. Accessed 20 Mar 2025

Teh WP et al (2022) DUB to the rescue. Mol Cell 82:1411–1413. https://doi.org/10.1016/j.molcel.2022.03.039. Accessed 20 Mar 2025
The 1000 Genomes Project Consortium (2015) A global reference for human genetic variation. Nature 526:68–74. https://doi.org/10.1038/nature15393. Accessed 20 Mar 2025
The All of Us Research Program Genomics Investigators (2024) Genomic data in the all of us research program. Nature 627:340–346. https://doi.org/10.1038/s41586-023-06957-x. Accessed 20 Mar 2025
The Cancer Genome Atlas Program (TCGA) (2025). https://www.cancer.gov/ccg/research/genome-sequencing/tcga. Accessed 20 Mar 2025
The Center for Protein Degradation at Dana-Farber Cancer Institute Boston, USA (2025). https://cpd.dana-farber.org/. Accessed 20 Mar 2025
The Centre for Protein Degradation ICR, London (2025). https://www.icr.ac.uk/research-and-discoveries/centres-and-strategic-collaborations/centre-for-protein-degradation. Accessed 20 Mar 2025
The Diaspora Human Genomics Institute (2025). http://thedhgi.org/. Accessed 20 Mar 2025
The wwPDB Consortium (2024) EMDB-the electron microscopy data Bank. Nucleic Acids Res 52: D456–D465
Theodoris CV et al (2023) Transfer learning enables predictions in network biology. Nature 618: 616–624. https://doi.org/10.1038/s41586-023-06139-9. Accessed 20 Mar 2025
TOPMed (2025). https://topmed.nhlbi.nih.gov/. Accessed 20 Mar 2025
Trost CN et al (2024) An anti-CRISPR that pulls apart a CRISPR–Cas complex. Nature 632:375–382. https://doi.org/10.1038/s41586-024-07642-3. Accessed 20 Mar 2025
Uhler C (2024) Building a two-way street between cell biology and machine learning. Nat Cell Biol 26:13–14. https://doi.org/10.1038/s41556-023-01279-6. Accessed 20 Mar 2025
UK Biobank (2023) World's largest genetic project opens the door to new era for treatments and cures: UK Biobank's major milestone. https://www.ukbiobank.ac.uk/learn-more-about-uk-biobank/news/world-s-largest-genetic-project-opens-the-door-to-new-era-for-treatments-and-cures-uk-biobank-s-major-milestone. Accessed 20 Mar 2025
Ultima Genomics Inc (2025). https://www.ultimagenomics.com/. Accessed 20 Mar 2025
Ünlü B et al (2023) Global analysis of suppressor mutations that rescue human genetic defects. Genome Med 15:78. https://doi.org/10.1186/s13073-023-01232-0. Accessed 20 Mar 2025
Van de Sande B et al (2023) Applications of single-cell RNA sequencing in drug discovery and development. Nat Rev Drug Discov 22:496–520. https://doi.org/10.1038/s41573-023-00688-4. Accessed 20 Mar 2025
Velimezi G et al (2018) Map of synthetic rescue interactions for the Fanconi anemia DNA repair pathway identifies USP48. Nat Commun 9:2280. https://doi.org/10.1038/s41467-018-04649-z. Accessed 20 Mar 2025
Võsa U et al (2021) Large-scale cis- and trans-eQTL analyses identify thousands of genetic loci and polygenic scores that regulate blood gene expression. Nat Genet 53:1300–1310. https://doi.org/10.1038/s41588-021-00913-z. Accessed 20 Mar 2025
Wang Z et al (2009) RNA-Seq: a revolutionary tool for transcriptomics. Nat Rev Genet 10:57–63. https://doi.org/10.1038/nrg2484. Accessed 20 Mar 2025
Wang Y et al (2021a) Nanopore sequencing technology, bioinformatics and applications. Nat Biotechnol 39:1348–1365. https://doi.org/10.1038/s41587-021-01108-x. Accessed 20 Mar 2025
Wang Q et al (2021b) Rare variant contribution to human disease in 281,104 UK biobank exomes. Nature 597:527–532. https://doi.org/10.1038/s41586-021-03855-y. Accessed 20 Mar 2025
Wang J et al (2022) Computational methods, databases and tools for synthetic lethality prediction. Brief Bioinform 23:bbac106. https://doi.org/10.1093/bib/bbac106. Accessed 20 Mar 2025
Wang Y et al (2023) Integrative splicing-quantitative-trait-locus analysis reveals risk loci for non-small-cell lung cancer. Am J Hum Genet 110:1574–1589. https://doi.org/10.1016/j.ajhg.2023.07.008. Accessed 20 Mar 2025

Wang Z et al (2024) Mendelian randomization analysis identifies druggable genes and drugs repurposing for chronic obstructive pulmonary disease. Front Cell Infect Microbiol:14. https://doi.org/10.3389/fcimb.2024.1386506. Accessed 20 Mar 2025

Wong C (2024) How to kill the 'zombie' cells that make you age. Nature 629:518–520. https://doi.org/10.1038/d41586-024-01370-4. Accessed 20 Mar 2025

Xiangdong LV et al (2023) Modulation of the proteostasis network promotes tumor resistance to oncogenic KRAS inhibitors. Science 381:eabn4180. https://doi.org/10.1126/science.abn4180. Accessed 20 Mar 2025

Yanai I et al (2024) Cellular atlases are unlocking the mysteries of the human body. Nature 635: 553–555. https://doi.org/10.1038/d41586-024-03552-6. Accessed 20 Mar 2025

Yang KD et al (2021) Multi-domain translation between single-cell imaging and sequencing data using autoencoders. Nat Commun 12:31. https://doi.org/10.1038/s41467-020-20249-2. Accessed 20 Mar 2025

Yang J et al (2023) Epigenetic regulation in the tumor microenvironment: molecular mechanisms and therapeutic targets. Sig Transduct Target Ther 8:210. https://doi.org/10.1038/s41392-023-01480-x. Accessed 20 Mar 2025

Yengo L et al (2022) A saturated map of common genetic variants associated with human height. Nature 610:704–712. https://doi.org/10.1038/s41586-022-05275-y. Accessed 20 Mar 2025

Yuan S et al (2024) Endoplasmic reticulum stress and therapeutic strategies in metabolic, neurodegenerative diseases and cancer. Mol Med 30:40. https://doi.org/10.1186/s10020-024-00808-9. Accessed 20 Mar 2025

Yun M et al (2011) Readers of histone modifications. Cell Res 21:564–578. https://doi.org/10.1038/cr.2011.42. Accessed 20 Mar 2025

Zanders ED (2016) Human drug targets a compendium for pharmaceutical discovery. Wiley Blackwell. https://www.wiley.com/en-ie/Human+Drug+Targets%3A+A+Compendium+for+Pharmaceutical+Discovery-p-9781118849828#evaluation-copy-section. Accessed 20 Mar 2025

Zhang D et al (2019) Metabolic regulation of gene expression by histone lactylation. Nature 574: 575–580. https://doi.org/10.1038/s41586-019-1678-1. Accessed 20 Mar 2025

Zhang P et al (2021) NyuWa genome resource: a deep whole-genome sequencing-based variation profile and reference panel for the Chinese population. Cell Rep 37:110017. https://doi.org/10.1016/j.celrep.2021.110017. Accessed 20 Mar 2025

Zhang S et al (2024) Protein neddylation and its role in health and diseases. Sig Transduct Target Ther 9:85. https://doi.org/10.1038/s41392-024-01800-9. Accessed 20 Mar 2025

Zhao S et al (2023) A single-cell massively parallel reporter assay detects cell-type-specific gene regulation. Nat Genet 55:346–354. https://doi.org/10.1038/s41588-022-01278-7. Accessed 20 Mar 2025

Zheng Y et al (2024) Precise genome-editing in human diseases: mechanisms, strategies and applications. Sig Transduct Target Ther 9:47. https://doi.org/10.1038/s41392-024-01750-2. Accessed 20 Mar 2025

Zhong Q et al (2023) Protein posttranslational modifications in health and diseases: functions, regulatory mechanisms, and therapeutic implications. Med Commun 4:e261. https://doi.org/10.1002/mco2.261. Accessed 20 Mar 2025

Zhong G et al (2024) Targeted protein degradation: advances in drug discovery and clinical practice. Sig Transduct Target Ther 9:308. https://doi.org/10.1038/s41392-024-02004-x. Accessed 20 Mar 2025

Zuo WF et al (2024) Heat shock proteins as hallmarks of cancer: insights from molecular mechanisms to therapeutic strategies. J Hematol Oncol 17:81. https://doi.org/10.1186/s13045-024-01601-1. Accessed 20 Mar 2025

Interventions

6

6.1 Introduction

Chapter 5 covered emerging approaches for discovering novel drug targets, while outlined here are some of the novel ways in which chemistry and biotechnology may be harnessed to intervene against these targets. This is where *expanding the landscape* from this book's title is most relevant, since in addition to small molecules, drug candidates include agents directed against the genome or its expression as well as therapeutics based on intact cells. The last two modalities have the potential to be curative after only a single administration (*one and done*), but gene and cell therapy is complex and expensive and therefore a burden on healthcare systems, particularly in the developing world. This is where, despite the need for repeat dosing over possibly many years, small molecule approaches have the advantage of a relatively low cost of goods and high patient compliance. Small molecules, protein, nucleic acid and cell-based therapeutics are covered in the sections below which focuses on the ways in which technical difficulties in discovery, clinical deployment and large-scale production might be overcome.

6.2 Small Molecules

Revisiting a comment in Chap. 4 about the properties of an ideal medicine: it should be orally bioavailable, induce disease remission, or at least long-term control in a wide patient population, be well tolerated and cost effective. The features that define therapeutically effective small molecule drugs[1] have been understood for decades. Despite the obvious successes of medicinal chemistry and pharmacology over the years the high failure rate in developing novel and effective compounds is a cause for

[1] i.e. with a molecular weight up to about 700 daltons.

E. D. Zanders, *Twenty-first Century Drug Discovery*,
https://doi.org/10.1007/978-3-031-93949-5_6

Table 6.1 The main issues arising in small molecule drug discovery

Features	Potential issues
Potency	*Undruggable* target proteins
Selectivity	Promiscuous binding
Metabolism and PK	Undesirable molecular features
Synthetic feasibility	Complex multistage synthesis
Manufacturing issues	e.g. Expensive and/or toxic reagents

concern. The reasons for failure are often well understood and are prompting new ways of dealing with the problems, the main ones being listed in Table 6.1.

6.2.1 Chemical Space

In a much-quoted paper by Bohacek (1996), chemical space contains a minimum of 10^{60} compounds with sizes and elemental composition consistent with small molecule drugs. It is unlikely therefore that only a minute fraction of these could ever be screened computationally (and certainly not synthesised), although perhaps quantum computing will one day be sufficiently developed to take on the first part of this challenge (see later comments). However at present, compound selection is a trial-and-error process often biassed by structures already used in existing drugs or screening libraries against target classes such as kinases for example. Furthermore, chemical groups known to be problematic due to pharmacokinetic, metabolic or toxicology issues will be excluded. A truly agnostic process would almost certainly reveal novel and effective chemotypes, but the question then is how might this be achieved? Medicinal chemistry and related areas are vast subjects, but it is possible to identify a few trends and issues relating to drug discovery as outlined in the following sections.

6.2.1.1 Flatland

Escape from Flatland: Increasing Saturation as an Approach to Improving Clinical Success. The title of this paper by Lovering et al. (2009) comes from the 1884 social satire *Flatland: a romance of many dimensions* by Edwin Abbott, in which the inhabitants of a two-dimensional world encounter higher orders of geometric space (Abbott 1884). Drug molecules interact with their (mostly) protein targets by changing 3-D conformations, which can favour aliphatic sp^3-hybridized structures over sp^2-hybridized aromatic molecules, hence the escape from the *flat land* occupied by the latter. However, aromatic ring systems are used extensively in small molecule drug discovery for reasons listed by Zhang et al. (2024):

Conformational restriction in binding sites is imposed by aromatic rings
Unique non-covalent interactions can enhance receptor binding
Addition of substituents to aromatic rings generates diverse compound collections

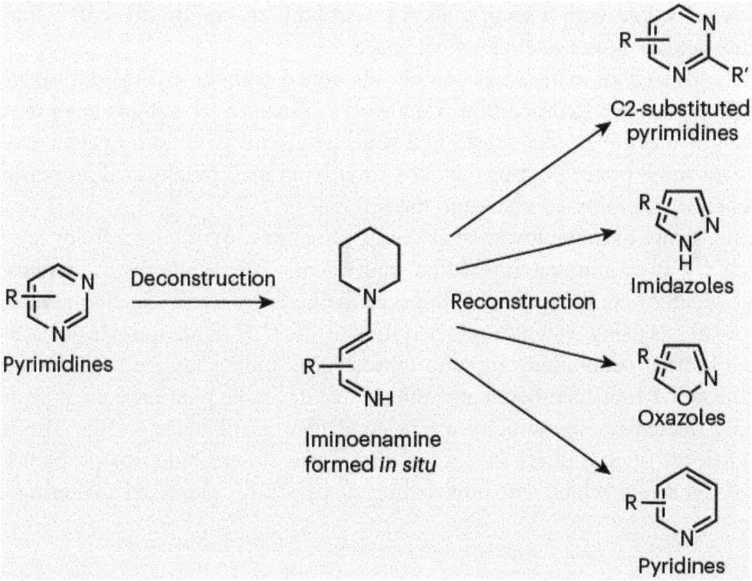

Fig. 6.1 Skeletal editing of pyrimidines producing different heteroarenes. (From Rentería-Gómez and Gutierrez (2024) with permission)

There are downsides to using aromatic scaffolds however, including metabolic instability, poor water solubility and increased lipophilicity leading to increased off-target effects. However, a significant proportion of FDA-approved drugs contain nitrogen heterocycles such as pyridine and benzannulated analogues since the basic framework can be functionalised by established synthetic techniques such as transition metal-catalysed cross coupling in a process called *skeletal editing*. A recent example is the conversion of pyrimidines into a number of different heteroarenes as illustrated in Fig. 6.1 (from Rentería-Gómez and Gutierrez 2024).
The exploration of chemical space for drug discovery covers unknown territory as well as that occupied by established pharmacophores. Some recent examples are illustrated in the following sections:

Many small-molecule drugs have been constructed by attaching chemical groups to arenes to form 2-D arrangements of groups that fit into the binding pockets of biological targets. In contrast to arenes, aliphatic molecules can bend and flex, allowing them more opportunities to occupy 3-D (rather than 2-D) space and fit snugly into target binding pockets. If cross-coupling methods could be developed to synthesize 3-D aliphatic molecules, scientists would be able to make compounds in the iterative drug-design process more quickly (Sorbie and Diao 2024). Al Zubaydi et al. (2024) have gone some way towards achieving this by using a novel Ni(alkyl) complex to couple with a second alkyl electrophile to rapidly diversify a range of structures and explore novel chemical space.

sp^2-hybridized aromatic rings can be substituted with sp^3-hybridised isosteres in the form of bicyclic hydrocarbon frameworks. Zhang et al. (2024) have used this approach to substitute meta-substituted benzene groups with nortricyclane isosteres in selected small molecule drugs thereby improving their biophysical properties and stereochemical activity compared to the original.

In an attempt to move towards sp^3 based structures, pyridine scaffolds are being replaced by the saturated piperidine equivalent. The synthetic challenges then become greater than for pyridines, but new methods are under development including biocatalysis using enzymes selected for specific C-H oxidation (He et al. 2024b).

Most nitrogen-containing rings in current drug molecules are five or six membered, however four-membered azetidines have desirable pharmaceutical properties including decreased lipophilicity and a good pharmacokinetic profile. The under-representation of azetidines in current drugs can be attributed to problems with directed synthesis which are now being successfully addressed (Wearing et al. 2024).

6.2.1.2 Natural Products

For many years the biopharmaceutical industry has had an on-off relationship with natural product discovery. The attractions are obvious given that many best-selling drugs have their origins in natural products and that the diversity of chemotypes found in nature far exceeds those made synthetically. Unfortunately identifying molecules of interest in adequate amounts to work with, let alone manufacture, can turn natural product chemistry into a major enterprise. The process of *diversity-oriented synthesis* (DOS) was introduced to create synthetically feasible compounds with the advantages of natural product functionality. The results have been disappointing however, perhaps due partly because 83% of the unique ring systems found in natural products are absent from synthetic libraries. A comparison between natural products and approved small molecule drugs shows on average that the latter have lower molecular weights and lipophilicities (logP), fewer chiral centres, rings and sp^3 carbons, but more aromatic rings (Kissman et al. 2024). A solution to the diversity and synthesis problems may come in part by exploiting the biosynthetic properties of enzymes, which after all, are responsible for making natural products in the first place. The biosynthetic genome encodes the blueprint for the synthesis of an enormous collection of small molecules but is mostly silent at any given moment. As a result, the full metabolic diversity of life remains enigmatic despite the rapidly growing wealth of sequence data and genome mining methods available. In addition to three-dimensional architecture and elaborate ring systems, many structures or functional groups previously considered to be abiotic, such as the nitro, nitrile, phosphonate or diazo groups, as well as elements such as fluorine, arsenic and selenium have also been found in secondary metabolites along with the enzyme families that install them (Kissman et al. 2024).

Enzymes can be used to produce chimaeric natural products or macromolecules that blend biological and synthetic attributes to access preparative routes or improve target binding and modification. The genetic encoding of biosynthetic pathways (genotype) and the synthesis of small molecule metabolites (phenotype) reflects the biological evolution of natural products serving different functions in the microbial world. However, as pointed out by Grigalunas et al. (2022) natural evolution itself is very slow and, in tandem with selection pressures, has resulted in natural products occupying only a fraction of the chemical space theoretically available. Different strategies have been used to expand chemical space, one of which, diversity-oriented synthesis, has been mentioned already. DNA-encoded libraries (DELs) provide another approach. DELs are created by covalently attaching small molecules to a unique DNA barcode that is PCR amplified and sequenced to identify target-binding ligands. Libraries with as many as 10^{12} members can be pooled together and analysed for target binding in a single tube. DEL technology and applications have been reviewed by Peterson and Liu (2023). Other design principles for chemical space exploration include the following (from Grigalunas et al. 2022):

Biology-Oriented Synthesis

Identification of simplified natural product core structures that may retain biologically relevant characteristics and decoration with substituents that simplify chemical synthesis while retaining biological activity. The diversity is limited since the core structures are produced from existing biochemical pathways.

Complexity to Diversity

Common natural product intermediates can undergo different biosynthetic pathways to arrive at chemically and biologically diverse end points. The complexity to diversity approach employs these latter molecules as divergent intermediates and chemically extends biosynthetic pathways by distorting the scaffolds of NPs through ring distortion reactions. The resulting compounds have diverse scaffolds distinct from the parent natural product but retaining the latter's complexity and biological relevance.

Pseudo-natural Products

Grigalunas et al. (2022) have developed the concept of *pseudo-natural products* to explore chemical space through fragment-based discovery. A pseudo-natural product can generally be defined as a synthetic or biosynthetically engineered small molecule that contains two or more natural product fragments in arrangements that are not found in nature. The authors have defined about 2000 natural product fragment groups that can be recombined to produce novel structures independent of the original natural product and that retain their chemical and biological relevance independently of known biosynthetic pathways. Their paper provides detailed examples of pseudo-natural products which occur unplanned in many compound collections. There is also a discussion of how evolution can be accelerated by engineering biosynthetic pathways from micro-organisms to produce novel compounds for example, by using the yeast synthetic biology platform described by Klein et al. (2014).

Figure 6.2 shows examples of the three categories for expanding chemical space in natural products described above by Grigalunas et al. (2022) as listed above.

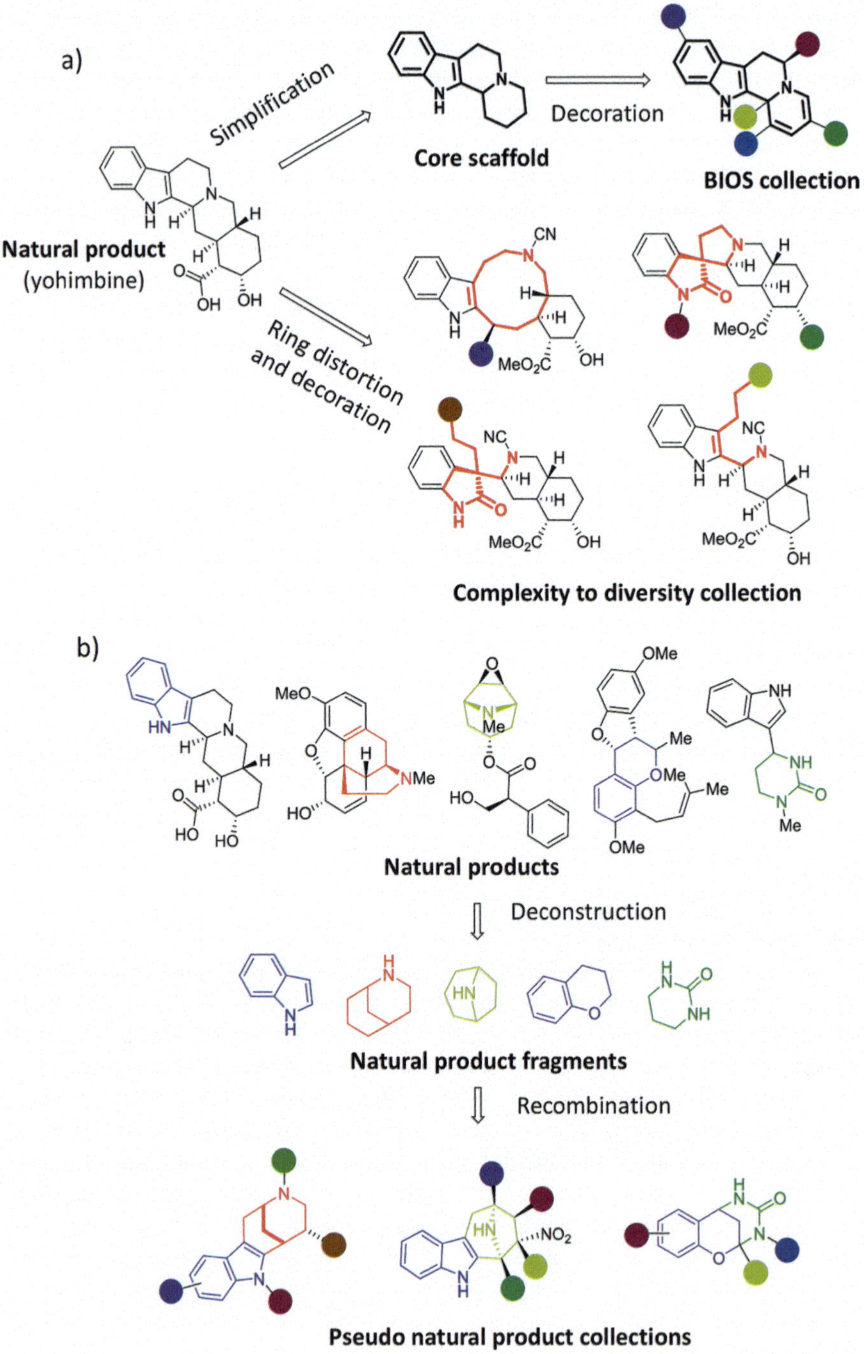

Fig. 6.2 Expanding natural product diversity. (Modified from Grigalunas et al. 2022 under CC-BY-4.0)

6.2.1.3 Machine Learning and Chemical Space

Machine learning and deep learning have been introduced in Chap. 3 with their algorithms being used to explore chemical space for drug candidate identification and selection. Activity in this area is intense, with tools being developed based on deep learning and associated algorithms, one example being DrugEx (Šícho et al. 2023). Physical compound libraries of course are used for screening, but they can only explore a small part of chemical space. Virtual on-demand databases such as REAL produced by Enamine (see chap. 3), increase the numbers by orders of magnitude. Generative spaces, unlike on-demand spaces, comprise theoretically possible molecules and collectively could comprise all chemical space with theoretical plausibility, estimated as 10^{23}–10^{60} drug-like compounds (Sadybekov and Katritch 2023).

Vogt (2023) has reviewed the issues involved in creating generative models for exploring chemical space, some of which are highlighted as follows:

Generative models randomly sample and explore the properties of chemical space. aiming to learn probability distributions from training data, which are then used to guide the generation of novel molecules.

Chemical space can be described algorithmically by using combinatorial approaches that systematically explore chemical structural graphs of increasing size. This has been achieved for molecules of limited size (up to 17 heavy atoms) and limited atom types (C, N, O, S, and halogens) the largest iteration (CDB-17) producing166 billion molecules. While impressive, this clearly falls far short of what is required for a comprehensive survey. The numbers could be substantially increased if arbitrary chemical structures (encoded as graphs and following basic valence rules) are produced at random and filtered to remove chemically nonsensical structures. However, the approach taken by many current deep learning approaches substitutes graphs with linear representations of molecules like SMILES (for graph and SMILES representations see Chap. 3). The probability of generating valid molecular representations is extremely low, but probabilities can be adjusted to focus on reasonable chemical structures. Deep neural networks can be used to explore chemical space using multiple parameters and not structural similarity alone. These parameters include synthetic accessibility, a quantitative estimate of drug-likeness and target-specific activity, all essential for drug development and covered below.

6.2.2 Selecting Candidates for Drug Development

Small molecules with activity against a chosen target must fulfil a number of criteria before entering the perilous phase of preclinical development (see Chaps. 2 and 7). The following sections outline some of these with an emphasis on virtual molecules that must be pre-screened to eliminate those with undesirable characteristics, thus saving a considerable amount of downstream time and effort.

6.2.2.1 Synthetic Feasibility

Digital Chemistry

Virtual compounds identified using computational methods may have desirable characteristics for exploring novel chemical space (sp^3 structures, low aromaticity etc) and yet be impossible to synthesise. The field of organic synthesis began in 1828 with the production of oxalic acid and urea by Wöhler. Since then, the *art* of chemical synthesis has been developed to a high degree of sophistication with challenging chemical targets.[2] Despite the introduction of computers for synthesis planning, chemists have been reluctant to move away from well-established methods but now it is inevitable that computer-aided synthetic design will be widely adopted for planning and executing syntheses. However, to achieve this computers need data input.

There is a recent trend for organic chemists to publish ever larger numbers of examples in methodology papers, but these reports remain focused on the knowledge and the dataset published in a journal article which represents only a small portion of the raw data collected. These data have not yet been collected in a standardized manner, and highly complex substrates are often not included. In more general terms, in the 200-year history of organic synthesis, we have not yet developed methods to collect, clean and label data in a way that makes it useful for training in the context of new reaction optimization, especially in the areas of catalysis design and development. Existing datasets in the public or private domain have simply not been built with this in mind. There is clearly a need for digital repositories along the lines of freely accessible DNA and protein sequence databases (Davies 2019).

Chemical synthesis data are essentially the *above the arrow* conditions marked in Fig. 6.3. and which for many projects could involve a complex and lengthy series of reagent additions and product purifications.

Machine learning is beginning to make inroads into this problem by, for example, identifying patterns from extensive reaction datasets to predict novel reaction outcomes. Reaction conditions can be optimised to maximise reaction yields and selectivity (Ali et al. 2024). Despite this promising scenario, challenges remain. It is difficult to interpret the decision-making processes used by deep learning algorithms (the *black box*) which is a problem, as chemists need to understand why specific predictions or recommendations are made; a deep understanding of chemical reaction mechanisms and complexity is crucial in this case.

Automation

Robotic chemistry is a reality but requires a consistent unified chemical language system for cross platform implementation of the chemical literature and adapting to new synthetic methods. There are many robotic systems, but these lack a common

[2] Out of 74 Chemistry Nobel prizes awarded since 1950, 11 have been for advances in chemical synthesis methodology.

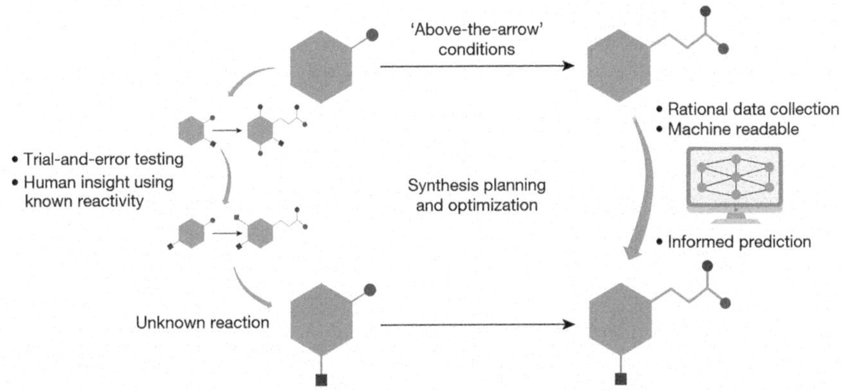

Fig. 6.3 Above the arrow conditions and digitisation of organic synthesis. (From Davies 2019 with permission)

architecture; because of this, automation in effect moves the work required from manual implementation to computer programming. Mehr et al. (2020) have developed a chemical programming language and its associated visual development environment that allows users to encode procedures without ambiguity and exchange them using XDL, a Chemical Description Language based on the ubiquitous XML format. Syntheses written in natural language are processed by an algorithm called SynthReader to allow digital input into XDL and implementation of chemical reactions across different platforms.

Another aspect of chemistry automation is the role of human decision making. Closed-loop autonomous chemical synthesis often bears little resemblance to human experimentation, either in the laboratory infrastructure required or in the decision-making steps. There is a danger in relying too much on machine learning algorithms pretrained with existing data, or adhering too rigidly to known rules, since novel reactions and/or products could be missed. This is one of the arguments described by Dai et al. (2024) who have used mobile robots in a modular workflow consisting of an automated synthesis platform and subsequent analysis by separate LC-MS and benchtop NMR instruments. A heuristic decision-maker[3] processes the measurement data, selects successful reactions to take forward and automatically checks the reproducibility of any screening hits.

6.2.2.2 Finding the Hits

Physical Screening

Compounds for small molecule discovery must be evaluated on target molecules (usually proteins) using physical or virtual screening. Any hits must then be progressed through a development programme to turn them into clinical candidates.

[3] i.e. *rule of thumb decision* making without analysing the full set of data.

The required screening technology such as binding or functional assays, or high content screening on whole cells, is well established for synthesised molecules. The level of throughput, i.e. low, medium, high and ultrahigh throughput formats will depend upon the specific assay required and the number of compounds available. The latter may come from random collections, focussed libraries, or fragments with low affinity binding to provide scaffolds for elaboration. No doubt there will be new developments for physical compound screening in the future, but the virtual chemical world is where real novelty is likely to occur as discussed here and in other parts of this chapter.

At this stage it is worth considering the off target or *promiscuous* binding of screened molecules (see Gilberg and Bajorath 2019). This has obvious implications for identifying *clean* compounds that selectively affect the chosen target, but also for considering binding to multiple targets in a beneficial way. The latter approach, known as *polypharmacology* could explain the therapeutic effects of some drugs which appear to go beyond its action on a single target (Bolognesi and Cavalli 2016). The challenge for drug developers is to identify true positives from screening campaigns and if desired, to deliberately create compounds that bind to more than one target.

Virtual Screening
Results generated by computer algorithms must correspond to the real world, be it 3-D structures of proteins, or small molecule hits from virtual screening. Mechanisms do exist for objectively comparing the output from different organisations, notably through the *Critical Assessment of Techniques for Protein Structure Prediction* (CASP) workshops that validated AlfaFold as a credible technology. Similarly, the *Critical Assessment of Computational Hit-finding Experiments* (CACHE) competition aims to do the same for small molecule binding to targets since there is intense activity in the field of virtual screening, with machine learning, or otherwise. The problem is to reliably finding potent binders that are not screening artefacts like those observed in wet lab screening. It is also a multiparameter optimisation problem, looking for compounds with the above features but also including good solubility and pharmacokinetic profiles etc.

The first programme (CACHE 13) was aimed at the WD40 repeat (WDR) domain of LRRK2, a multifunctional enzyme associated with Parkinson's disease. Twenty three teams predicted over 2000 candidates for binding, but fewer than twelve of these actually fit into the WDR pocket (with binding affinities ranging from 20 to 70 micromolar), less than a 1% hit rate (see Mullard 2024a for details). It is fair to say that the LRRK2 target was chosen as a challenging *undruggable* target, but the exercise is clearly important for revealing the best computational techniques for this kind of problem.

It is impossible here to comprehensively review the field of screening small molecules (both real, virtual, synthetic or natural product) particularly in the era of machine learning and large datasets. Instead, some references to ongoing approaches

Table 6.2 Selected references in computational compound screening and drug design

Summary	Reference
Machine learning and drug design review	Turzo et al. (2022)
Deep learning and transcriptomics	Aliper et al. (2016)
Proteome-wide drug MOA	Mitchell et al. (2023)
Large-scale ligand discovery using chemoproteomics	Offensperger et al. (2024)
ChemBert a-2, deep learning multiparameter optimisation	Rollins et al. (2024)
Minimum variance sampling analysis: True and false positives in HTS data	Boldini et al. (2024)
Machine learning and natural product discovery	Mullowney et al. (2023)
Antibiotic discovery with explainable deep learning	Wong et al. (2024)
Deep learning and QSAR	Tropsha et al. (2024)
Chemistry42	Ivanenkov et al. (2023)

are presented in Table 6.2. These include using mRNA and protein expression data to map the effects of compounds on targets, the application of machine and deep learning to antibiotic and other natural product discovery, drug design and QSAR (quantitative structure activity relationships).

6.2.2.3 Quantum Computing and Virtual Screening

The demands on computing hardware and software imposed by massive datasets and complex algorithms are increasing relentlessly. This is as true for drug discovery as it is for many other aspects of life, so the excitement generated by advances in quantum computing, highlighted in Chap. 3, seems warranted. Some aspects of quantum computing in chemistry have been discussed by Service (2024). Advances in quantum computing hardware have been made to the point where companies such as PASQAL, Qubit Pharmaceuticals and Gero are beginning to evaluate the technology on drug discovery problems. Since the quantum computing technology is at an early stage, a hybrid classical/quantum approach is being employed (for example Gircha et al. 2023) albeit with the expectation that future computing will be purely quantum-based. Vakili et al. (2025) have used the hybrid method to find potential KRAS inhibitors[4] through screening one million virtual compounds in a generative model that used a combination of a 16-qubit quantum computer and conventional machine learning on cloud computers, along with filtering for synthesizability and pharmacological properties. Fifteen promising ligands were identified of which two showed broad activity across KRAS mutants; results from this hybrid computing approach indicated superiority over fully classical models, and introduced potential chemotypes that expanded the diversity of KRAS inhibitor design. This is seen as a promising start towards the goal of achieving rapid and comprehensive small molecule designs through pure quantum computing on larger machines.

[4] See also Sect. 6.2.3.

6.2.3 Drugging the Undruggable

A major objective of small molecule discovery is to expand the range of targets from druggable (i.e. with high affinity binding sites) to the undruggable, i.e. those lacking suitable binding pockets. Despite decades of searching for potent small molecules against the latter type, the limited success in this area has driven the uptake of biotherapeutics such as monoclonal antibodies. However, recent years have seen the discovery of small molecule clinical candidates against targets that had previously been considered to be undruggable. This section covers the general approach for finding small molecules that bind to seemingly *flat* protein targets as well as indirect approaches related to target degradation. A successful small molecule approach would remove the need for complex and expensive therapies based on proteins, nucleic acids and cells, thus reducing healthcare costs, and treating both common and rare diseases. Although finding small molecules with equal or greater clinical efficacy to biotherapeutics may lie in the future, there are examples of discovery programmes aimed in this direction. For example sickle cell anaemia is treatable with gene therapies (see later); however, small molecule and antibody approaches are at various stages of development, some of which inhibit the gene therapy target and others that inhibit haemoglobin association in red blood cells, or the critical enzyme, pyruvate kinase (reviewed by Sheridan 2024).

6.2.3.1 Covalent Inhibitors: The Example of KRAS

Most small molecule drugs act as reversible inhibitors or activators; on the other hand, covalent molecules, permanently modify their targets, acetylsalicylic acid (aspirin) being a prime example. Other examples include penicillin, omeprazole, 5-fluorouracil and bortezimib (Raouf 2024). Despite these successes, for many years covalent binding was considered to be an undesirable characteristic due to non-specific reactivity and unpredictability of responses. There are clear advantages however, as covalent drugs can achieve picomolar potencies and minimal interference by competing substrates. They also display persistent *in vivo* activity beyond metabolic clearance, which extends the systemic half-life and reduces dosing. From the point of view of tackling undruggable targets, covalent inhibitors allow the exploration of challenging binding pockets; this opens the way for a new generation of molecules, of which KRAS inhibitors are notable examples.

The oncogene Kirsten rat sarcoma viral homolog (KRAS) integrates upstream receptor tyrosine kinase (RTK) signalling with downstream signals to regulate cell growth. KRAS protein is a GTPase that cycles between a protein *on* state (GTP bound) and an *off* state (GDP-bound). Oncogenic mutations in KRAS shift the balance toward a perpetual *on* state and unchecked growth, a situation occurring in many cancers. KRAS was an uncrackable safe, having picomolar affinity for GTP and no obvious allosteric sites (Herzberg and Manji 2023). A breakthrough came with the discovery of a covalent allosteric inhibitor which held mutant KRAS in an inactive form (Ostrem et al. 2013). The particular mutation (G12C) presented a reactive cysteine residue that could be covalently targeted; building out with

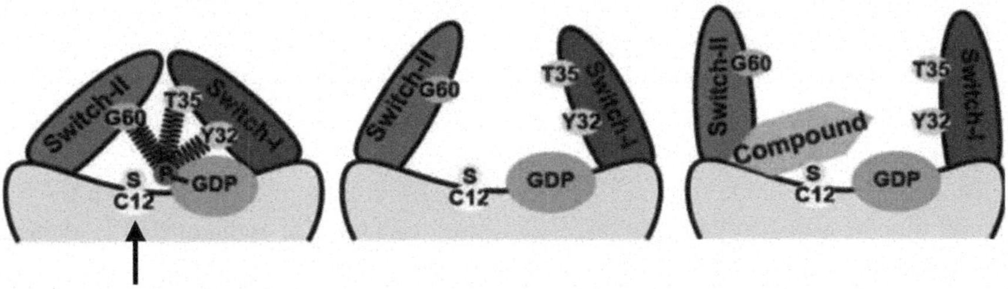

Fig. 6.4 Covalent inhibition of KRAS *on* state by targeting mutant C12 residue (arrowed). (Modified from Ostrem et al. (2013) with permission)

fragments revealed a hidden pocket adjacent to the GDP binding site that could be used to lock KRAS(GDP) in an inactive state (see Fig. 6.4.).

Approved clinically active KRAS inhibitors (e.g. adagrasib) are now available having established the basic principle of covalently binding to mutant residues such as cysteine. This has been extended to other mutations like G12S (covalent inhibitors) and G12D (non-covalent inhibitors) (Herzberg and Manji 2023).

Medicinal chemistry approaches for covalent drugs usually include the use of weakly reactive electrophiles such as acrylamide, or reversible covalent chemistries including nitriles, aldehydes, boronic acids, α-ketoamides, and α-cyanoacrylamides (Raouf 2024).

6.2.3.2 Indirect Targeting: Protein Degradation

If a protein target cannot be directly influenced with a small drug-like molecule, one option is to remove it altogether using targeted protein degradation. This aspect of proteostasis involving the ubiquitin proteosome system has already been covered in Chap. 5. but examples of compounds developed as target degraders are given here.

Firstly, medicinal chemistry is directed towards the discovery of new molecular glues and PROTACS (Ishida and Ciulli 2021). Early PROTACs were bispecific molecules consisting of a ligand binding the target of interest and an E3 ligase-binding moiety separated by a synthetic linker. This arrangement is not ideal, as the ligand for the E3 ligase adds complexity in the form of higher molecular weight and lipophilicity, thus compromising oral bioavailabilty (Hughes et al. 2021). Despite these issues, and the spectre of drug resistance, effective new therapies based on the principles of targeted protein degradation will almost certainly be developed. The following examples give an indication of some current activity in the field:

Sickle Cell Disease

As briefly mentioned earlier, the search for small molecule alternatives to gene therapy for treating this disease has concentrated on controlling haemoglobin aggregation or aspects of red blood cell physiology. There is also ongoing work to stimulate haemoglobin F (HbF) production as a protective mechanism, one that also forms the basis of gene therapy (see later). Ting et al. (2024) have exploited molecular glues to achieve this using a small molecule degrader of the WIZ transcription factor that represses HbF production. A chemical library biassed towards the E3 ligase cereblon (CRBN) was used to screen for compounds that induced HbF in an erthyroblast assay. A promising candidate dWIZ-1 was found to recruit the WIZ transcription factor to cereblon leading to its degradation and upregulation of HbF. The compound has been further evaluated *in vivo* in humanised mice and cynomolgous monkeys. Although the compound was discovered by serendipity in a phenotypic screen, the desired outcome of inducing HbF clearly vindicated the choice of molecular glue mechanisms as a way of addressing this problem.

KRAS Degradation

Mutant KRAS oncogenes were once seen as undruggable targets, but as described earlier have been effectively targeted with covalently (and non-covalently)- bound small molecules. Another approach involves targeted protein degradation as described by Popow et al. (2024) who have identified a binding pocket that allows for reversible inhibition of a number of different KRAS mutants, thus broadening the range of tumours that could be targeted. The VHL E3 ligase was selected for a medicinal chemistry approach using a suitable linker to the KRAS ligand. The resulting PROTAC molecule was rigorously tested and shown to inhibit a number of different KRAS mutants with tenfold greater potency than direct inhibition. The structure had to be optimised for improved pharmacokinetics in mouse *in vivo* studies, but the observed anti-tumour responses for G12D and G12V mutants showed the potential of this pan-KRAS degradation approach in oncology.

HIV-1 Nef

The HIV-1 accessory protein Nef an attractive antiretroviral drug target as it enhances viral infectivity and promotes immune escape. Nef lacks any known biochemical or enzymatic activities, functioning instead through interactions with a diverse range of host cell proteins at the plasma membrane, trans-Golgi network and other membranous sites. While this creates difficulties with small molecule inhibition of Nef function, it is possible to find modest binders, as described by Emert-Sedlak et al. (2022). The same authors have taken this further by linking Nef-binding compounds to cereblon (CRBN) and Von Hippel–Lindau (VHL) E3 ligases (Emert-Sedlak et al. 2024). Eight out of twelve PROTAC candidates degraded Nef and had the desired biological effect of restoring MHC I and CD4 expression in virally infected T-cells thus providing further opportunities for treating AIDS by clearing HIV-1 infection.

6.2.4 Selected Issues in Small Molecule Drug Discovery

Some important selected issues in small molecule drug discovery and development are considered in the following sections.

6.2.4.1 Target Classes

The earlier sections have not focussed specifically on any small molecule drug target classes such as enzymes, receptors, transporters, or ion channels; however, an exception is made here for G-protein coupled receptors (GPCRs) because of their potential for producing even more useful medicines than is already the case. In addition there is a short section on RNA as a target for small molecules, something not normally considered because of biopharma's emphasis on proteins.

G-Protein Coupled Receptors (GPCRs)

The GPCR family consists of about 800 members, roughly half of which are odorant or taste receptors. To emphasise their fundamental importance in drug development, over 135 GPCRs are targets for roughly 35% of FDA approved drugs. Many of their natural ligands are small molecules and therefore druggable; the problem lies with the membrane association of the seven transmembrane GPCR proteins, along with different signalling complexes, making structure-based design challenging in the absence of reliable 3-D structures. Luckily the introduction of cryoelectron microscopy, membrane solubilisation and other techniques has moved the field along considerably as highlighted by Bob Lefkowtiz, one of the pioneers[5] (Mullard 2023). Apart from the need to find synthetic (or even natural) ligands for so-called orphan GPCRs, there is a need for more selective drugs, particularly against families binding the same ligand. For example, there are 15 different serotonin (5-HT) receptors with very similar ligand binding (*orthosteric*) sites. Given the significant biological effects of individual 5-HT receptors, particularly in the brain, some *allosteric* binding is called for where the sites are distinct on different receptors.

Biased signalling results from different pathways being activated by the same receptor and could be exploited to separate desirable from undesirable responses elicited by a given ligand. For example, it may be possible to separate the analgesic effects of drugs acting at κ-opioid receptors (through G-protein signalling) from respiratory depression (through β-arrestin signalling). The two aspects of fine tuning GPCR responses, namely allosteric binding and biased signalling are discussed in detail by Kenakin (2024) and illustrated in Fig. 6.5.

RNA Targets for Small Molecules

About 2% of the genome encodes protein, yet 80% of the rest is transcribed and regulated as noncoding RNA. Most of the latter is of unknown function and roughly 85–90% of protein targets are largely undruggable (Gilman quoted in Garber 2023). The estimate for protein druggablilty may be somewhat pessimistic, but the point is made about the potential of RNA as a drug target in its own right. While research activity is at an early stage, three main areas are being pursued, namely RNA processing, RNA splicing and RNA degradation (see Garber 2023 for more background). This area is of interest to biopharma, particularly in light of the success of splicing modifiers (covered below) and the early-stage discovery companies that focus exclusively on this area, some of which are listed in Table 6.3.

[5] And Nobel Prize recipient along with Brian Kobilka in 2012.

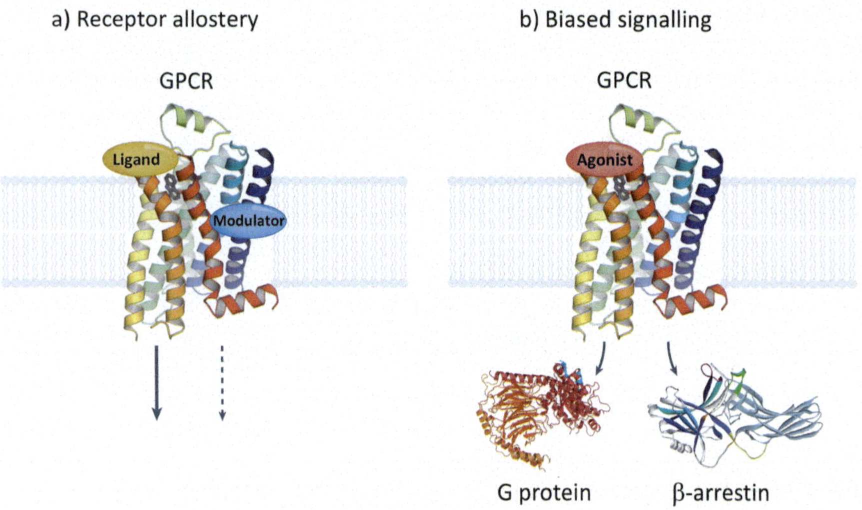

Fig. 6.5 Modulating GPCR signalling through allosteric binding (**a**) and biased signalling (**b**). (Modified from Kenakin (2024) with permission)

Table 6.3 Some early-stage companies focused on RNA as a drug target

Atomic AI	Machine learning for RNA structures
Arrakis therapeutics	RNA-targeted small molecules
Expansion therapeutics	RNA processing
Remix therapeutics	RNA processing
Ribometrix	RNA-targeted small molecules
Skyhawk therapeutics	RNA splicing

RNA splicing modifiers were developed by PTC Therapeutics as a way of correcting the aberrant splicing in spinal muscular atrophy (SMA) that leads to defective SMN1 protein. The small molecule drug Evrysdi® (risdiplam) was discovered by PTC Therapeutics in a functional screen for splicing activity and developed by Roche. The loss of functional SMN1 protein can be compensated for by production of SMN2 which is normally absent due to lack of exon 7 in the mRNA. Evrysdi stabilizes the interaction between the 5′ splice site of exon 7 and the U1 snRNP resulting in enhancing splicing, recovery of exon 7 and production of functional SMN2 protein. A simplified diagram of the process is shown in Fig. 6.6.

The proof of concept afforded by the small molecules affecting pre-mRNA splicing is encouraging work on other targets, for example in Huntington's disease characterised by elevated levels of mutant huntingtin protein (HTT). It is possible to induce splicing into the mRNA of a pseudoexon containing a premature stop codon, thus preventing the production of the full length HTT protein and rapidly degrading the mRNA.[6]

[6] Achieved by Novartis with bramaplam but discontinued. Active programmes by PTC Therapeutics and others.

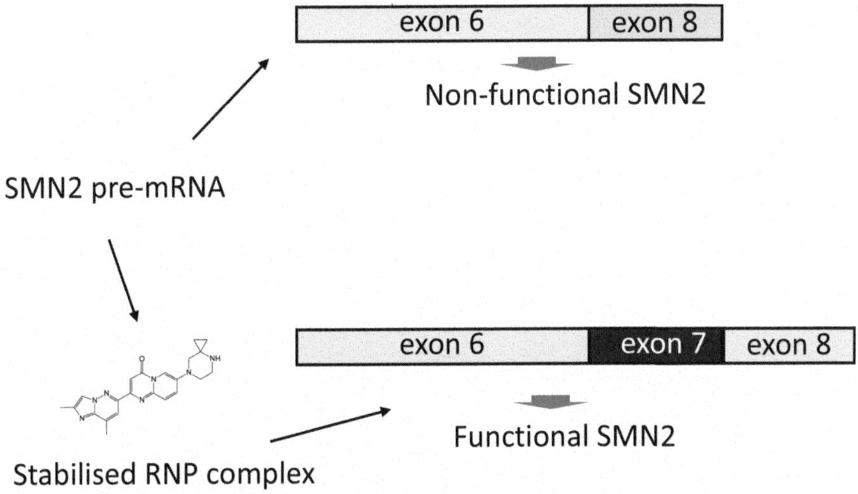

Fig. 6.6 Overcoming splice inhibition in SMN2 mRNA

Targeted degradation of mRNA using RIBOTACs, by analogy with PROTACS, is another possible approach to exploiting RNA with small molecules. Tong et al. (2023) have created a ribonuclease-targeting chimera in which an RNA-binding molecule is appended to a heterocycle that binds to and locally activates, RNase L1. As proof of concept, selective degradation was demonstrated for microRNA-155 JUN mRNA and MYC mRNA.

6.2.4.2 Early Development

This section covers two issues encountered in early development of small molecules, namely oral bioavailabilty and improved pharmacokinetics.

Orally Bioavailable Peptidic Macrocycles

Macrocycles are ring structures found in many complex natural products and as such have the potential to bridge the gap between conventional small molecule drugs and larger peptidic structures. Their size and chemical properties make them unlikely candidates as orally absorbed drug molecules; however, this has been achieved by Merck in the form of MK-0616 a 1500 dalton engineered macrocycle targeted to proprotein convertase subtilisin/kexin type 9 (PCSK9) a genetically validated drug target (Kingwell 2023). The structure of MK-0616 (now International non-proprietary name (INN) *enlicitide*) is shown in Fig. 6.7. PCSK9 binds to the LDL receptor (LDLR) on the surface of hepatocytes to trigger internalization and degradation of the receptor thereby affecting uptake of LDL-C from the circulation. Receptor degradation can be inhibited by blocking the PCSK9-LDLR interaction an enabling LDLR recycling which reduces plasma LDL-C levels, a risk factor for coronary disease. Different approaches to inhibiting PCSK-9 are in development and the clinic, with the ideal being an orally available molecule. MK-0616 is of obvious

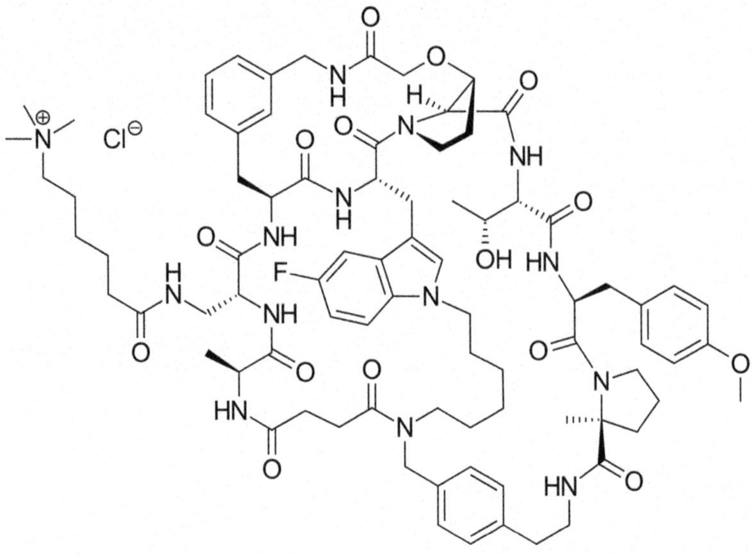

Fig. 6.7 Structure of MK-0616 (Enlicitide)

interest for that reason, but also because of the fact that it can be made orally available at all. This gives hope that the medicinal chemistry approaches used by Merck to achieve this can be applied to other macrocycle drugs.

Merck collaborated with Ra Pharma (now UCB) by using its ExtremeDiversity mRNA display system to generate trillions of peptide macrocycles for screening against the PCSK9 target (and others). The hit macrocycle named 44 was optimised by replacing a thiol-based linker and central triazole for stability and changing the northern olefin to an amide crosslinker to increase solubility (via reduction in lipophilicity) and to avoid isomers generated from the olefin formation; the resulting compound MK-0616 had a potency of 5pM (Johns et al. 2023). The molecule was too large and polar to be absorbed though cell membranes, so it was formulated with sodium caprate which acts as a permeation enhancer by temporarily loosening the tight junctions between intestinal epithelial cells.

Although the above macrocycle is unable to cross the cell membrane, the well-studied immunophilin cyclosporin A (CYPA) can achieve this by changing its conformation in lipophilic and aqueous environments before binding to the intracellular protein calcineurin. Revolution Medicines (see publication by Schulze et al. 2023) exploited this mechanism by using a cyclosporin-binding molecule (sanglifehrin A) to simultaneously bind KRAS using a cysteine-seeking warhead for covalent binding, and CYPA with high affinity. The resulting bifunctional molecule was active *in vitro* and *in vivo* as well as being orally bioavailable. This illustrates the promise of macrolide drugs as agents for influencing previously

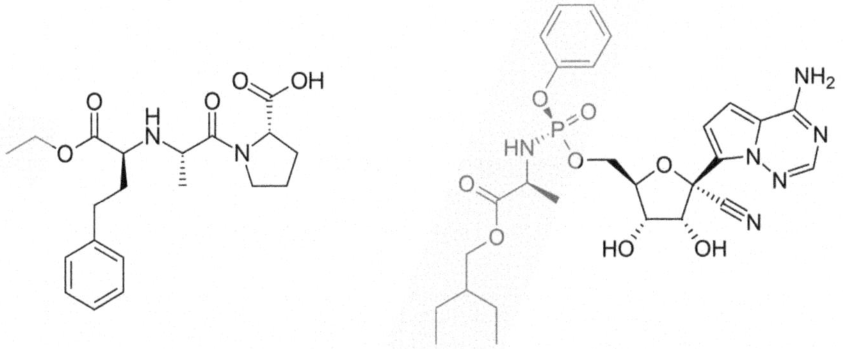

Enalapril Remdesivir

Fig. 6.8 Examples of two prodrugs with cleaved moieties shown in grey

undruggable targets while conferring oral bioavailabilty. Other activities in the macrolide area are discussed by Kingwell (2023).

Prodrugs

Prodrugs are modified active pharmaceutical ingredients (APIs) that are released during biotransformation *in vivo*. They can be exploited to enhance drug delivery (oral and parenteral), increase cell and tissue permeation and duration of action, and reduce toxicity. Two examples are given in Fig. 6.8. the angiotensin converting enzyme (ACE) inhibitor enanapril and the SARS-Cov-2 RNA polymerase inhibitor remdesivir. Enanapril is converted into the active drug enanaprilat by the action of esterases which is also the case for remsdesivir, although the latter undergoes further biotransformations into the active nucleoside triphosphate molecule.

The field has been reviewed in detail by Fralish et al. (2024) who have collated data on 178 FDA-approved prodrugs including 197 biotransformation mechanisms[7] as shown in Table 6.4.

The prodrug field is mature, but there are opportunities for developing novel strategies, for example in targeting individual species of commensal bacteria with antibiotics without affecting the microbiome in general. To achieve this, Evans et al. (2019) have synthesised an antibiotic prodrug that combines ciprofloxacin with a β-lactamase-cleavable motif that is only bactericidal after activation by β-lactamase, an enzyme absent in human cells and only present in a subset of bacteria.

Other examples of prodrug design for particular therapeutic situations are given in Fralish et al. (2024).

[7] Some drugs use more than one biotransformation.

Table 6.4 Biotransformation mechanisms for 178 FDA-approved prodrugs

Mechanism	%
Esterases	37
Phosphatases	7
Other hydrolases	6
Oxidoreductases (CYP450)	17 (12)
Kinases	17
Non-enzymatic biotransformations	6
Non-human nitroreductases	4

Data from Fralish et al. (2024) with permission

Prodrug design is potentially well suited to predicting ADMET profiles, routes of metabolism and associated toxicities through machine learning and could be a viable alternative to other drug delivery strategies. Prodrug approaches add more complexity compared with conventional drug design since safety evaluation is necessary for the prodrug, the drug itself, and all associated metabolites, but given the number of effective prodrugs in the clinic this has clearly not been an insurmountable problem.

Deuterium in Small Molecule Drugs

Substitution of hydrogen atoms in a drug molecule with deuterium (deuteration) is a subtle change, but one which can have an impact on its pharmacokinetic or toxicity profile (see also Chap 2. and Di Martino 2023). The C–D bond is more stable than the C–H bond due to the twofold increase in the mass of D over H and the slower reaction kinetics (the *kinetic isotope effect*). Initial deuteration studies modified existing drugs to improve their pharmacokinetics, a process known as *deuterium switching*. In 2017, the FDA approved deutetrabenazine, a deuterated form of the vesicular monoamine transporter 2 inhibitor tetrabenazine, the former having a greatly superior pharmacokinetic profile which allowed for a significant reduction in dose and dosing frequency. Deuterium substitution is now used at the outset of a discovery programme rather than just on an existing medicine. The TYK2 kinase inhibitor deucravacitinib, approved by the FDA in 2022, is the first example of a novel deuterated drug; deuterium substitution avoids the formation of a non-selective metabolite and preserves the exquisite specificity of the parent drug for TYK2 over other enzymes. There is now a trend towards producing deuterated drug candidates *de novo* rather than through deuterium switching; detailed studies of the way in which deuteration affects drug metabolism provides opportunities for creating medicines with good PK profiles. There are problems of course, including the production of deuterium-containing molecules at high purity, but the area in general provides yet another example of the relentless advance of small molecule drug discovery.

6.3 Proteins

Sales of protein biotherapeutics now account for over 50% of the world's top ten selling branded medicines yet are expensive to develop and manufacture and require parenteral administration. Nevertheless as drugs, proteins are indispensable for treating disease in the absence of suitable small molecule equivalents; the combination of undruggable small molecule drug targets, the exquisite specificity of proteins and the ability to recruit the immune system to execute functions such as targeted cell killing means that protein therapeutics will be produced by the biopharmaceutical industry for many years to come. Monoclonal antibodies currently provide the mainstay of protein therapies used for a autoimmune, allergic, infectious and neoplastic diseases. Antibodies are also being developed for other conditions, for example addiction to the synthetic opioid fentanyl (Bremer et al. 2023).

The immunoglobulin molecule allows high affinity binding to target antigens along with recruitment of the immune system via Fc receptors; its basic structure, however, can be altered through genetic engineering to provide novel functionalities as illustrated in Fig. 6.9. and highlighted below:

- Radiolabelled Antibodies

Radionuclides directed to tumour cells by monoclonal antibodies induce local apoptosis and necrosis thereby sparing bystander tissues. The radioactive element chosen can be either alpha, beta, or Auger electron emitters. Beta emitters are the most frequently used, having a long range and low energy transfer, examples including lutetium-177 (^{177}Lu), yttrium-90 (^{90}Y), and iodine-131 (^{131}I) (Parakh

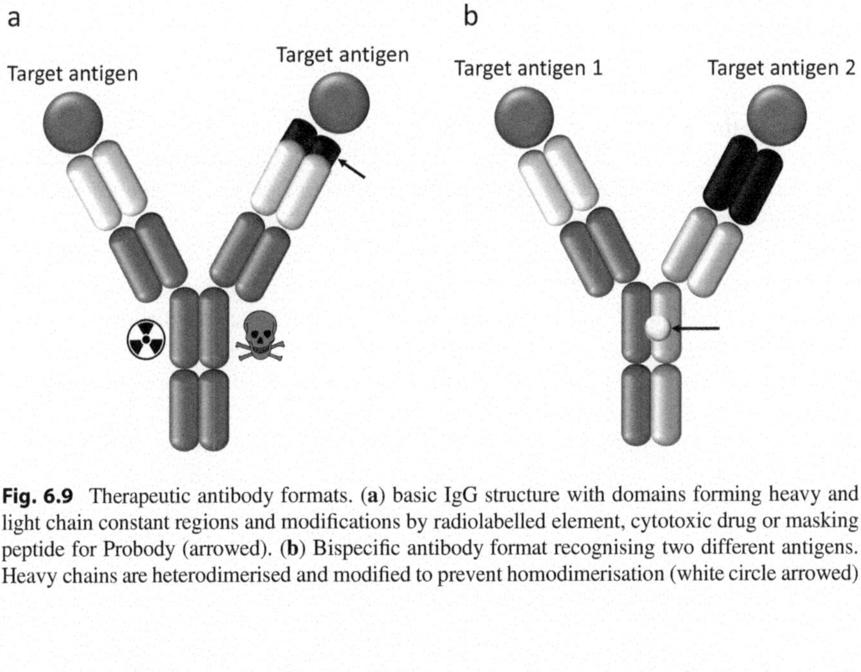

Fig. 6.9 Therapeutic antibody formats. (**a**) basic IgG structure with domains forming heavy and light chain constant regions and modifications by radiolabelled element, cytotoxic drug or masking peptide for Probody (arrowed). (**b**) Bispecific antibody format recognising two different antigens. Heavy chains are heterodimerised and modified to prevent homodimerisation (white circle arrowed)

et al. 2020). Alpha particles have shorter path lengths with higher energy emission thereby increasing tumour cell killing with reduced bystander effects. Alpha emitters include astatine-211 (^{211}At), actinium-225 (^{225}Ac), thorium-227 (^{227}Th), and bismuth-213 (^{213}Bi). Clinical experience with radioactive antibodies has been gained over many years, but they appear to be most effective on haematological tumours, with solid tumour therapy being limited though toxicity. Future strategies could include targeting patients with minimal disease burden or using newer radionuclides (including alpha-particle isotopes), combined therapeutic modalities (such as DNA sensitizers), or locoregional application (Parakh et al. 2020).

- Antibody-drug Conjugates (ADCs)

ADCs deliver cytotoxic payloads to a target cell (normally a tumour). The conjugated drug payload must be 100- to 1000-fold more effective than the free drug since only a very small fraction of the ADC reaches the target. A number of issues need to be considered, including antibody affinity, the nature of the linker (cleavable or non-cleavable) and the cytotoxic payload. ADCs are entering the clinic and show promise for cancer treatment in particular (e.g. Dumontet et al. 2023). Future work with ADCs will require the optimisation of different parameters such as pharmacokinetics and efficacy, the stability of the cytotoxic payload and the use of different protein constructs such as antibody prodrugs (*probody therapeutics*) or *nanobodies* (see later).

- Degrader-Antibody Conjugates (DACs)

Targeted protein degradation using the ubiquitin-proteosome system has been covered earlier in connection with intracellular small molecule targets. However, many drug targets reside on the cell surface or are extracellular, a situation ideally suited for antibody-directed approaches. For example, the cytotoxic drug in an ADC can be replaced with a molecular glue-based protein degrader that allows for precise targeting to the relevant cell (News feature Nat.Biotech (2023); Nurix). It would also be useful to couple antibodies and targeted degradation within an intracellular environment. For example, Benn et al. (2024) have removed intracellular tau protein complexes as a strategy for treating neurodegenerative diseases such as Alzheimer's. The E3 ubiquitin ligase tripartite motif-containing protein 21 (TRIM21) contains a RING domain that activates E3 ligase activity upon protein crosslinking. The combination of TRIM21 and tau-specific antibodies can clear extracellular, but not intracellular tau. Benn et al. (2024) engineered the intracellular expression of a nanobody fused to the TRIM21 RING domain resulting in the desired outcome of the selective depletion of intracellular tau aggregates.

In addition to the ubiquitin-proteosome system of targeted protein degradation, there are other biological mechanisms available for use in engineered constructs, for example the removal of cell surface mucin glycoproteins from the surface of tumour cells. Pedram et al. (2023) linked a mucin-specific bacterial metalloprotease to a

Table 6.5 Receptors used for lysosome-targeting chimaeras (LYTACS)

Cation independent mannose-6-phosphate receptor (CI-M6PR)
Asialoglycoprotein receptor (ASGPR)
Macrophage galactose type lectin 1 (MGL1)
Integrins
Scavenger receptors (SRs)
Cytokine receptors

nanobody recognising the HER2 tumour antigen resulting in the death of $HER2^{+}$ and not $HER2^{-}$ cells *in vitro*.

- Lysosome-Targeting Receptors (LTRs)

Novel strategies are being developed for extracellular targets that use various cell surface receptors and the lysosome pathway (reviewed by Chen et al. (2023). Some of these *lysosome targeting receptors* (LTRs) exploited in *lysosome-targeting chimaeras* (LYTACS) are listed in Table 6.5).

Some LYTAC based therapeutics are entering the clinic (Mullard 2024b). Biohaven have developed BHV-1300 a bispecific LYTAC that targets circulating IgG antibodies[8] and traffics them specifically to the liver via the asialoglycoprotein receptor (ASGPR). The small molecule triantennerary *N*-acetylgalactosamine (tri-GalNAc) is a ligand for ASGPR and can be used as part of a bifunctional molecule. A phase 1 clinical trial showed up to 43% reduction in circulating IgG, thus supporting the LYTAC concept. Other companies investigating this area include Lycia Therapeutics, Avilar Therapeutics and EpiBiologics.

While the above LYTAC approaches are promising, monoclonal antibodies can remove extracellular antigens via the lysosomal pathway without modification. Extracellular binding occurs at neutral pH, but internalisation of the antigen/antibody complex can lead to dissociation in the endosome at low pH thereby removing the extracellular protein and recycling the antibody to reengage target antigens. This phenomenon can also be exploited to create a *sweeping antibody* that has pH dependent antigen binding as well as increased binding to the cell surface neonatal Fc receptor (FcRn) at neutral pH (Igawa et al. 2013).

- Polyspecific Antibodies

Antibodies engineered to simultaneously recognise more than one antigen represent the next generation of therapies based on monoclonal antibodies. The technology required for engineering multiple antigen-binding structures into the basic immunoglobulin molecule (or variant) is extensive as reviewed by Sandeep et al. (2024). Most of the current biopharma activity involves constructs recognising just two or three different antigens. Bispecific antibodies (BsAbs) are engineered to recognise two separate antigens, for example one on the surface of a tumour cell and the other

[8] To treat autoimmune diseases by removing autoantibodies from the circulation.

the CD3 molecule on a T lymphocyte. The objective is to draw the cells together in close proximity to allow efficient tumour cell killing. The field has been reviewed in detail by Herrera et al. (2024) who describe various BsAb configurations, and the strategies used to assemble the different H and L chains needed for the full-sized immunoglobulin.

The basic immunoglobulin structure can also be engineered to preventing Fc crosslinking thereby minimising the risk of cytokine release syndrome. The pharmacokinetics of IgG can be improved by extending its serum half-life by engineering binding to the neonatal Fc receptor (FcRn) on vascular endothelial cells. The binding internalises the antibody to protect it from catabolism and then recycles it to the cell surface (Akilesh et al. 2007).

Trispecific antibodies incorporate an extra antigen binding element to enhance the level of immune cell activation achieved with bispecific antibodies. Promsote et al. (2023) have generated a trispecific antibody that recognises the envelope protein of the HIV virus, CD3 as the second antigen and separately CD28 on CD4 and CD8+ve T cells. The antibody construct drives the release of latent HIV virus from infected cells followed by its elimination by CD8 T cells activated by CD28 costimulation. This is a potential means of eliminating the virus from T cell reservoirs, which is one of the main challenges in completely eliminating the HIV virus in AIDS patients.

The tetraspecific GNC-035 protein developed by Systimmune Inc. represents a further expansion of polyclonality: GNC-035 recognises a cancer target ROR-1, along with CD3 on T lymphocytes to induce cytotoxicity, PDL1 to stop checkpoint inhibition and 41-BB on accessory cells to enhance T-cell activation.

- Antibody Prodrugs: Probody Therapeutics

Antibody prodrugs work on the same principle as their small molecule analogues: inactive when administered but converted into an active entity through biotransformation. Probody therapeutics exploit the elevated levels of proteolytic enzymes within the tumour microenvironment such that a Probody can enter the circulation without causing immune related adverse reactions and release the active antibody within the tumour. The molecule is built from an immunoglobulin specific for a tumour antigen, a masking peptide linked to the N-terminus of the light chain, and a protease-cleavable substrate linker peptide. The technology has been applied to different antibody formats including immunotherapies, Probody drug conjugates, and T-cell-redirecting bispecific Probody therapeutics (Autio et al. 2020).

- Antibodies for Difficult Targets

Difficult targets refer to challenging small molecule targets such as GPCRs and ion channels where limited binding site availability or lack of selectivity can be a problem. Evolution has provided many natural products in the form of the toxins and venoms that directly interfere with ion channels and GPCRs and which have historically provided starting points for medicinal chemistry and drug development. Spider venoms are a rich source of disulfide-rich peptides, including a class of inhibitor cystine knot (ICK) peptides called *knottins*, so-called because they have a disulphide-rich structural motif that forms a *knot* conferring high structural, thermal and proteolytic stability (Dongol et al. 2019). Spider venom knottins target voltage-gated sodium channels (NaV) involved in pain sensation and therefore a focus of interest in the development of novel analgesics (see also Chap 4).

One way of targeting the appropriate knottin peptide to the correct ion channel subtype is to incorporate the sequence into an antibody CDR domain. The resulting KnotBodies™ are being developed by Maxion Therapeutics for blocking selected ion channels.

6.3.1 Protein and Peptide Engineering

Protein engineering using recombinant DNA technology has been used since the pioneering days of biotechnology in the 1970s. Modifications to natural proteins like antibodies and protein hormones have, and are, being made to control biological activity and improve pharmacokinetics, stability and other practical issues for drug therapy, some examples of which are covered below including the development of machine learning approaches.

6.3.1.1 Cytokines

The large family of cytokines, their receptors and signalling pathways are becoming well understood after decades of research. Although small molecule drugs affecting the ligand receptor binding component have proved elusive, monoclonal antibodies are effective and are represented in a significant number of top selling biotherapeutics. The field has been extensively reviewed by Leonard and Lin (2023) who highlight the toxicity problems found with interleukin-2 (IL-2) administration in the clinic. IL-2 acts by binding to trimeric high-affinity IL-2 receptors formed of α (CD25), β (CD122) and γ (CD132) subunits. CD25 confers high affinity binding while the intermediate affinity receptor IL-2Rβγ, consists of β- and γ-subunits that mediate signal transduction. High-dose IL-2 (aldesleukin) has been approved as first line immunotherapy for patients with metastatic renal cell carcinoma and malignant melanoma. However, its anti-tumor immunity can be compromised by the induction of regulatory T-cells (Tregs) and activation-induced cell death. In addition, systemic IL-2 treatment has been associated with severe cardiovascular, pulmonary, hepatic, gastrointestinal, neurologic and haematological side effects. A group at Roche have engineered an IL-2 variant that does not bind CD25 but still signals through the βγ subunits, thus avoiding the above-mentioned toxicity issues (Klein et al. 2017). The resulting immunocytokine (cergutuzumab amunaleukin) consist of a single IL-2 variant (IL2v) moiety fused to the C-terminus of a high affinity bivalent carcinoembryonic antigen (CEA)-specific antibody devoid of Fc-mediated effector functions. The design avoids preferential activation of regulatory T-cells vs. immune effector cells, increases the therapeutic index of IL-2 therapy and improves the pharmacokinetics (Klein et al. 2017).

The above example illustrates the general approach which has been applied to other aspects of immunotherapy, including a built-in PD-1 binding module linked in *cis* to the IL2 variant lacking CD25 binding (Codarri Deak et al. 2022).

6.3.1.2 Antibodies

Clinical experience with the first monoclonal antibodies showed that antibody engineering was essential to convert rodent immunoglobulins into proteins with human-derived amino acid sequences. Following on from the first chimaeric antibody constructs, fully humanised monoclonals are now produced routinely using a variety of techniques. The popularity of monoclonal antibodies as biotherapeutics is very clear but there is clearly room for improvement whether using the basic immunoglobulin structure, or the various modifications covered earlier. One of the main problems is the size of the molecule, as is clear from Fig. 6.10 which shows the size ranges from small molecules via engineered peptides and small proteins up to the 150 kDa immunoglobulin G molecule. The *bicycle* and *nanobody* constructs represent an attempt to retain the binding affinity and specificity of the IgG molecule within a smaller format that can improve pharmacokinetics and manufacturability.

Nanobodies: Single Domain-Based VHHs

Antibodies from dromedary camels and other camelids (eg. llamas and alpacas) are devoid of the CH1 heavy chain domains needed for light chain pairing; as a result they are truncated heavy chain only proteins (Jin et al. 2023). Camelid antibodies have a molecular weight of 90 kDa which can be reduced by removal of the Fc region to a VHH fragment of 15 kDa which has full antigen binding capability. Their

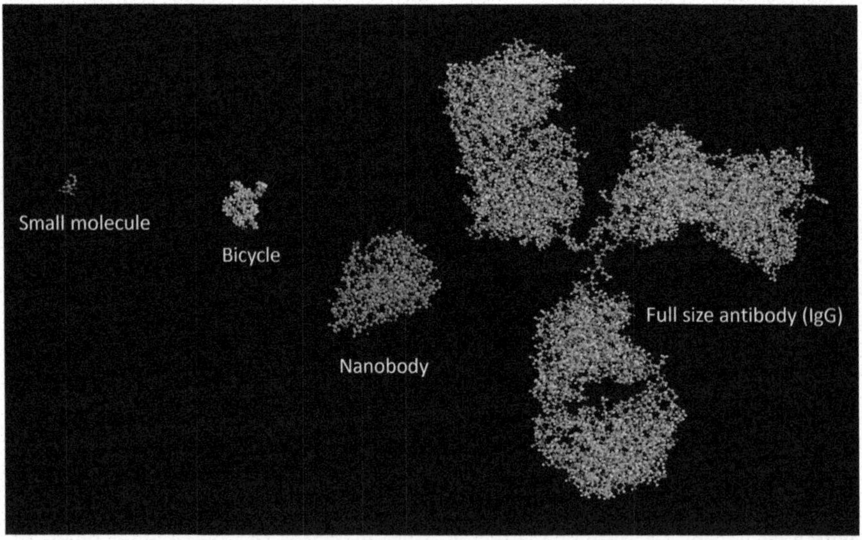

Fig. 6.10 Size comparison between immunoglobulins and other drug types. (Images from RCSB PDB)

small size and 3-D structure confers thermal stability and protease resistance allowing oral administration (Hussack et al. 2011). So far, three nanobodies have been approved as therapeutic drugs, while ongoing and future work includes designing multivalent VHH fragments that preserve the binding affinity seen with conventional immunoglobulins and can be conjugated with radiolabels or cytotoxic drugs (Jin et al. 2023).

Bicyclic Peptides

Peptidic macrocycles have already been described in connection with oral delivery of small molecules lacking classical drug-like properties. As the ring size of monocyclic peptides increases however, they become more conformationally flexible and susceptible to proteolytic degradation; stability can be improved by creating bicyclic structures, thus reducing the size of each ring to approximately half of the original size and increasing conformational rigidity. Furthermore, the bicyclic system permits the two rings to function independently of each other, e.g., one ring for target binding while the other for cellular entry (Rhodes and Pei 2017). Bicyclic peptides are naturally occurring but can also be produced synthetically using mRNA display or phage display technologies. Bicycle Therapeutics has pioneered the use of phage display to create a range of therapeutic candidates with more than 10^{20} linear peptides being expressed by phage libraries and reacted with scaffolds to form bicycle molecules on the phage surface. The company is able to configure over 300 different library formats thus providing significant diversity for screening programmes. Bicycle molecules display good tissue penetration as a result of their small size (1.5–2.5 kDa), which makes them attractive candidates for treating solid tumours. They can also be produced using chemical synthesis which is another advantage over conventional antibodies.

6.3.1.3 Machine Learning and Protein Design

The success of AlfaFold in predicting 3-D protein structures and functions from primary amino acid sequences using deep learning approaches was highlighted in Chap.5. The inverse problem of predicting a primary sequence from a desired function has been addressed by David Baker, co-winner of 2024 Nobel prize for Chemistry along with the inventers of AlphaFold. His RoseTTAfold software can design protein shapes to order using databases of short amino acid sequences with useful structural motifs. The basic principles have been described by Chu et al. (2024) and Notin et al. (2024). Nagra et al. (2023) have published an industry analysis of activity in the protein design field which has been energised by the success of algorithms such as AlphaFold and RoseTTAfold. The basic elements of computational protein design are illustrated in Fig. 6.11 based on the description by Chu et al. (2024).

One obvious application for computational design is in the field of protein therapeutics, specifically for antibody optimisation and the creation of novel proteins with a desired biological activity.

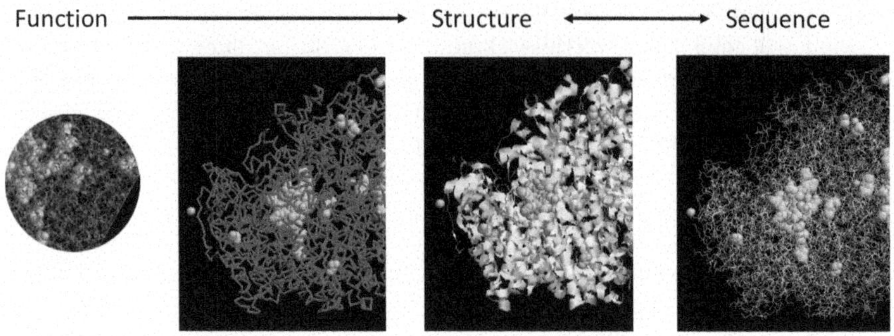

Fig. 6.11 Protein Design Strategy based on Chu et al. (2024). Images left to right: a protein structure is built around a functional element, such as ligand binding site or enzyme active site, and structures progressively developed until resolving individual amino acid sequences

Antibodies

Protein sequence data can be regarded as a language and therefore amenable to natural language processing. He et al. (2024a) list several protein language models that have been used to optimise antibody affinity, in one case with up to 160-fold improvement. Despite such progress, the field of antibody optimisation is limited by the great diversity of antibodies and the scarcity of antigen-antibody pairing data. This problem is being addressed in a number of ways; for example, Hie et al. (2024) have developed a pretrained generative large language model (PALM-H3) to generate high-affinity antibody CDRH3 sequences targeting a specific antigen, and A2binder for pairing antigen epitope sequences with antibody sequences to predict the binding specificity and affinity. They showed significant sequence differences between antibodies generated by PALM-H3 and natural antibodies, but binding was not significantly affected. CDRH3 sequences can vary in length for different antigens but the PALM-H3 model learns to perform insertions and deletions to generate useful variable-length CDRH3 outputs during the sequence-to-sequence generation process. It also captures important features that contribute to binding affinity having been trained on a large dataset of natural antibodies.

Another approach, employed by Shanker et al. (2024), uses native protein backbone structures to guide an inverse folding protein language model towards identifying novel sequences without the need to use sequence-based training data. Their results support the notion that the protein's structure itself may be used in lieu of learned surrogate functions of evolutionary fitness and eliminates the reliance on any initial data collection or prior knowledge of the protein.

Cytokine Mimetics

Protein engineering of cytokines has already been discussed, but the *de novo* design of proteins for high affinity binding to cytokine (or other) receptors has generally failed for targets with relatively flat and polar surfaces. One of these is the tumour necrosis factor receptor 1 (TNFR1) which along with other superfamily members, are important drug targets. However, Glögl et al. (2024) have used a generative artificial intelligence method RFdiffusion that generates new protein structures by progressively denoising random spatial distributions of residues. Binding proteins can be generated for a target of interest without any guidance from preexisting scaffolds and are not biased toward any particular scaffold or limited by the number of available residues. The algorithm could generate folds with extensive shape matching to the target and enable the design of high-affinity binders (pM affinity) to targets for which previous design efforts had failed.

As a general point, new computational techniques mean that the search for high-affinity and high-specificity binders to therapeutically important and structurally challenging protein targets does not necessarily require animal immunisation, library design, or large-scale screening.

6.3.1.4 Hacking the Genetic Code

Proteins sequences are specified by 64 triplet codons used to direct ribosomal synthesis in a template directed manner. Apart from a few atypical examples, all proteins utilise 20 amino acids in various combinations. The genetic code is decoded using transfer RNAs (tRNAs) aminoacylated with their cognate amino acids by aminoacyl-tRNA synthetase enzymes. Efforts are ongoing to reprogramme organisms to incorporate noncanonical amino acids (ncAAs) into biopolymers by re-engineering (hacking) the translational machinery and recoding entire genomes (Chin 2017). Genetic code expansion uses an orthogonal aminoacyl-tRNA synthetase (aaRS)-tRNA pair to direct the incorporation of a ncAA into proteins, using an unassigned codon (commonly the amber stop codon, UAG) introduced at the desired site in a gene of interest. The orthogonal synthetase specifically aminoacylates an orthogonal cognate tRNA to the exclusion of endogenous synthetases. Although the UAG stop codon is commonly used, the production of truncated proteins can be a problem, so other strategies using rare codons are being developed (Ding et al. 2024).

There are clearly many biopharmaceutical applications for proteins modified with ncAAs including diagnostics, cell biology and the introduction of post-translational modifications. The latter can be used to engineer biotherapeutics for different applications, including:

Antibody drug conjugates: e.g. Incorporation of *p*-acetylphenylalanine into anti-Her2 antibody Fab fragments and full-length IgG to selectively and efficiently conjugate an auristatin derivative via a stable oxime linkage (Axup et al. 2012).

Modified recombinant proteins: e.g. growth hormone modified with *p*-acetylphenylalanine to allow derivatisation with polyethylene glycol (PEGylation) to improve pharmacodynamics (Cho et al. 2011).

Reprogrammed cytokine specificity: e.g. modifying interleukin-2 to interact with β and γ receptors only. This goal of removing the α chain specificity of the IL-2 receptor (IL-2R) has already been mentioned; the work of Ptacin et al. (2021) show how this result can also be achieved by changing the genetic code to incorporate N6-(2-azidoethoxy)-carbonyl-l-lysine (AzK) and creating a library of IL-2 variants which were then PEGylated using click chemistry. Variants could then be selected by screening for those with reduced IL-2 Rα engagement while still retaining IL-2Rβγ potency.

6.4 Nucleic Acids

The relevance of DNA and RNA to drug target discovery has been covered in Chap. 5. The nucleic acids themselves or their derivatives can be developed as therapies in their own right as *genomic medicines*. The overall picture of nucleic acid targets is presented in Fig. 6.12 with antisense, siRNA and aptamer drug molecules included.

Genomic medicines have gone through periods of hype and disillusionment since their initial conception with a significant number of clinical trial failures due to safety concerns. Xie et al. (2023) have analysed programmes for virally delivered gene therapies, genome editing technologies (e.g. CRISPR), antisense oligonucleotides and small interfering RNA (siRNA) therapies and reported that out of ~1100 active clinical trials of genomic medicines, only about 20 products were approved at the time. Of significance is the fact that the treatments target <0.1% of the global population with genetic disorders and 53% of programmes overall are for rare diseases. According to Xie et al. (2023), the landscape for genomic medicine development over the next few decades is likely to move from rare monogenic diseases to more common diseases as illustrated in Table 6.6. The actual outcome of course will depend upon scientific advances, including drug targeting and commercial aspects such as pricing, which are both mentioned later.

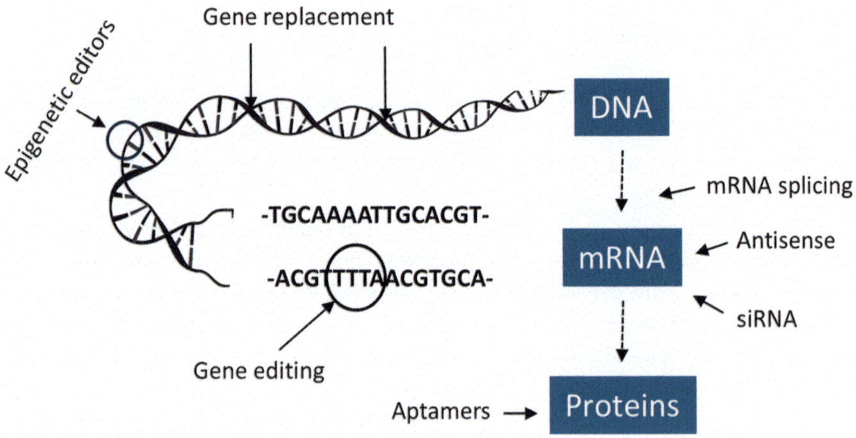

Fig. 6.12 Nucleic acid-based targets and therapeutics

Table 6.6 Possible waves of genomic medicine development over decades

Wave 1	Wave 2	Wave 3
Monogenic rare diseases of the eye, liver, haematopoietic system and CNS e.g. haemophilia, sickle cell disease and spinal muscular atrophy	Rare diseases such as Duchenne muscular dystrophy and cystic fibrosis, and non-rare diseases such as hypercholesterolaemia	Complex multigenic conditions, Potentially involving multiple organs e.g. rare diseases such as neuromyelitis optica, and non-rare diseases such as Alzheimer disease and various cancers

Adapted from Xie et al. (2023) with permission

6.4.1 Gene Therapy

Gene therapy involves manipulation of genomic DNA in patients, or cells derived from them, in order to replace faulty DNA sequences or to produce small edits. The technology is now quite mature, with gene editors CRISPR Cas9, Transcription Activator-Like Effector Nucleases (TALENs) and Zinc Finger Nucleases used for research and in the clinic. *Base editors* use nCas9 a catalytically impaired (dead) Cas9 enzyme to introduce a single-strand break on the guide RNA-targeted strand to establish the introduced mutation rather than repair it (Komor et al. 2016).

Prime editing combines CRISPR-Cas9 with reverse transcriptase to synthesise a single-strand DNA sequence from an RNA template and insert it into the DNA with the advantage that it edits bases without causing double-stranded DNA breaks. Prime Medicine are conducting the first clinical evaluation of Prime Editing for the treatment of $p47^{phox}$ Chronic Granulomatous Disease (CGD) caused by the production of abnormal white blood cells due to a mutation in NCF1. Another gene targeting approach uses the ARCUS[9] nuclease, a synthetic enzyme derived from the homing endonuclease I-CreI that shows limited off-target effects. Developed by Precision Biosciences, the enzyme system has been employed by iECURE to treat an infant with ornithine transcarbamylase (OTC) deficiency by delivering a corrected DNA sequence to the liver. The integration site is a so-called *safe harbour region* of the genome, in this case the PCSK9 gene which does not trigger the inappropriate gene expression that has been observed with other gene therapies. At the time of writing there have been anecdotal reports of clinical success, but no publications as yet (Kaiser 2025).

An alternative gene therapy approach involves rewriting the epigenome to alter gene expression with no or minimal changes in the underlying DNA sequence (Ueda et al. 2023). This is potentially suited to genetic diseases with gain-of-function genes, or for nongenetic diseases where the target gene can be increased by epigenetic modification. In practice, targeting sequences can be fused with epigenetic effector sequences to create EpiEffectors. One example is *coupled histone tail for autoinhibition release of methyltransferase* (CHARM) a compact, enzyme-free epigenetic editor that can recruit an endogenous DNA methyltransferase, induce

[9] From Latin, *Arcus*: bow.

hypermethylation of a target gene promoter, and durably silence expression of the gene and thus production of its protein product in the brains of mice upon delivery via a single AAV vector (commentary by Whittaker and Musunuru 2024). The CHARM system silences the Prnp promoter by targeting specific DNA sequences with ZFs coupled to D3L and an H3 tail, which recruit and activate endogenous DNMT3A. This reduces prion protein (PrP) expression in the brains of mice thus providing a strategy for treating neurodegeneration in prion diseases.

Examples of approved gene therapy medicines:

Hemgenix, (CSL Behring)
Hemgenix is a recombinant adeno-associated virus (rAAV) vector-based gene therapy for Haemophilia B, an inherited deficiency of coagulation factor IX that leads to an increased tendency to bleed. The disease is rare (1 in 20,000 to 50,000 live male newborns) but inadequately treated. The vector containing the Factor IX gene is produced in insect cells and introduced to the liver using a non-replicating recombinant adeno-associated virus serotype 5 (AAV5); this contains a codon-optimised cDNA of the human coagulation Factor IX variant gene under the control of a liver-specific promoter (LP1) (Hemgenix EMA data).

Casgevy, *Exagamglogene Autotemcel* (CRISPR Therapeutics, Vertex Pharmaceuticals)
This is an example of the first CRISPR Cas gene-edited product to be approved for clinical use. The target diseases are the haemogobinopathies sickle cell disease and β-thalassaemia containing defective adult haemoglobin. Therapeutic strategies involve restoring normal function by substituting the adult protein with correctly functioning foetal haemoglobin. Foetal haemoglobin production is suppressed during development through the action of the BCL11A transcription factor; the latter can be removed by editing the erythroid enhancer region of this gene thus removing the suppression of the foetal protein in adults (see Frangoul et al. 2024 for background and recent clinical data). In practice, the patient's CD34 + ve haematopoietic stem and progenitor cells (HSPCs) are modified *ex vivo* with the BCL11A guide RNA and cas9 protein by electroporation and reintroduced to the patient to allow the development of genetically modified erythrocytes. The process is complex and involved, as illustrated schematically in Fig. 6.13 and makes the point that gene therapies of this type are unlikely to address a large patient population and market size.

6.4.1.1 Gene Delivery

Viral Vectors
A suitable gene therapy vector for use in humans must be long lasting, targeted to the correct tissue(s), be engineered to avoid replicating, cost effective to manufacture and above all, be safe. The approved gene therapies described earlier used lentiviruses to infect stem cells *ex vivo* and adeno associated virus (AAV) for direct administration to the patient.

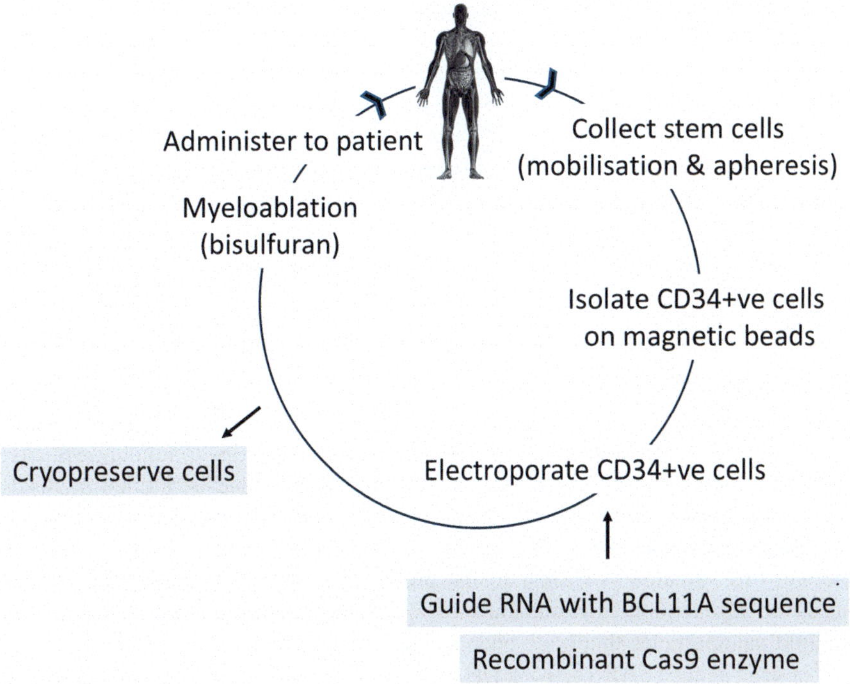

Fig. 6.13 Reinfusion of HPSCs after modification with CRISPR Cas9

Retroviruses are classified as α-, β-, γ-, δ- and ε-Retroviridae, Spumaviridae and Lentiviridae and contain an RNA genome which is reverse-transcribed into double-stranded DNA prior to being stably integrated into the host cell genome. First generation gene therapy vectors were derived from γ-Retroviridae but the trend has been towards using lentivirus vectors derived from HIV (Poletti and Mavilio 2021). In order to create a safe non-infectious HIV derived vector, all viral genes are replaced with a gene expression cassette containing only the sequences necessary for vector packaging, reverse transcription and integration.

Adeno associated viruses (AAV) (parvoviruses) have been used as gene therapy vectors since 1995. Recombinant AAV (rAAV) is a protein-based nanoparticle that can traverse the cell membrane and deliver the transgene DNA payload into the nucleus. It lacks the *rep* and *cap* genes of wild-type AAV and can be engineered to alter tissue tropism and evade host immunity (Muhuri et al. 2022). One of its attractions as a gene therapy vector lies in the fact that transgene expression can last for years due to the formation and maintenance of circular episomes in non-dividing cells. Despite the successful implementation of gene therapy using vectors of this type there are a number of issues that have to be addressed before the technology can be used more widely; some of this is covered in Muhuri et al. (2022) and in the detailed review of AAV gene therapy provided by Wang et al. (2024). For example, innate and adaptive immune responses to the viral protein must be blocked through pre-screening or administration of immunosuppressive drugs; in addition, the rAAV capsid has limited capacity for DNA inserts (about 4.7Kb) which precludes the packaging of large genes, for example that encoding the 14Kb mRNA for dystrophin. It is possible to package parts of the dystrophin gene in separate AAVs, but this is an inefficient process leading to scrambled proteins; however advances are being made in correcting the splicing of long mRNAs or proteins. For example, Lindley et al. (2024) have used ribozyme-activated mRNA ligation to correctly splice two separate mRNA sections of model genes, while Tasfaout et al. (2024) have engineered a natural protein *trans*-splicing mechanism into separate AAV constructs resulting in the correct assembly of full-length dystrophin protein in situ.

Non-viral Delivery

Great strides have been made with viral vectors for delivering nucleic acids to cells, but as mentioned earlier, there are potential safety issues relating to immunogenicity, viral replication through recombination, insertional mutagenesis, limited payloads and manufacturing costs. Ideally gene therapy would be *one and done*, but the inserted gene can wane over time leading to the need for further doses that may be rejected; this situation is similar to the problems found with repeat injections of murine or chimaeric monoclonal antibodies. A possible solution may be found with the lipid nanoparticles that have been so successful in delivering mRNA vaccines used at scale against SARS-Cov-2. DNA delivery without viruses has been reviewed by Sheridan (2023), with RNA delivery being covered later in the chapter. Any DNA delivery system must avoid recognition by the innate immune system and target the cells of interest, which often means having to bypass the liver. Circulating serum apolipoprotein E (Apo E) binds nanoparticles and transports them to ApoE receptors in the liver, so changes in nanoparticle formulations can reduce binding and direct the DNA or RNA cargo elsewhere. As an example, the correction of cystic fibrosis using gene therapy with the cystic fibrosis transmembrane conductance regulator (CFTR) requires gene delivery to the lungs, so Sun et al. (2024) have used specially designed lipid nanoparticles (LNPs) to deliver gene editing (base editor) molecules to murine pulmonary stem cells leading to a long-lasting correction in this animal model.

A number of delivery technologies like these are in early development and should result in gene therapy becoming safer, more flexible and cheaper in the future.

Some companies developing non-viral nucleic acid delivery systems are listed in Table 6.7 (see references for links).

6.4.1.2 Commercial Aspects of Gene Therapy

Gene therapies are expensive, as highlighted in Table 6.8 (adapted from Mullard 2024c).

The costs have to be borne by healthcare providers resulting, as in the case of Zynteglo, with the product being withdrawn from the market through pricing restrictions. Some of the conditions successfully treated with recent gene therapies,

Table 6.7 Companies developing virus-free DNA delivery systems

Company	Technology
Aera Therapeutics	DNA delivery with exosomes
Avectas	Delivery by permeabilising cell membranes
Code Bio	3D DNA synthetic DNA delivery platform
Generation Bio	ceDNA for nuclease resistance and nuclear targeting
Nanite	Polymer-based nanoparticles
Recode Therapeutics	Organ-targeted lipid nanoparticles

ceDNA closed ended DNA

Table 6.8 Gene therapy approvals and list cost

Name	Delivery	Cost ($M)	Year approved
Luxturna	AAV	0.85	2017
Zolgensma	AAV	2.1	2019
Zynteglo	Lentivirus ex vivo	2.8	2022
Skysona	Lentivirus ex vivo	3.0	2022
Hemgenix	AAV	3.5	2022
Vyjuvek	HSV topical	0.6305 p/a	2023
Elevidys	AAV	3.2	2023
Roctavian	AAV	2.9	2023
Lyfgenia	Lentivirus ex vivo	3.1	2023
Lenmeldy	Lentivirus ex vivo	4.25	2024

Adapted from Mullard (2024c) with permission

like sickle cell disease or thalassaemia, are more prevalent in the developing world where expensive therapies may be unaffordable. There is pressure therefore on manufacturers to reduce the cost of the therapy through pricing agreements such as value-based pricing. If gene (and cell) therapies are really going to be part of the future, the costs involved will have to go down through innovation at different stages of the development process. Rouce and Porteus (2024) have written a succinct review of these issues, pointing out that cell and gene therapies are expensive because they are custom-made with manufacturing of small batches or even single doses and therefore lacking the economies of scale required to make goods available to global populations. Supply chain costs can account for up to 50% of total manufacturing costs, but it is possible to reduce them. For example, analytical reagents with specifications required for use in humans can cost 10 to 100 times more than those with a *research use only* label while performing no better.

Innovation in clinical care can have a major effect; India has decreased the cost of haematopoietic stem cell transplantation by 30- to 50-fold compared to the USA while maintaining excellent clinical results. The removal of products from sale because of pricing issues could be addressed by commercial innovations such as public-private partnerships and not-for-profit foundations etc. These ideas have been explored in more detail by Harlow and Adair (2024) and Kliegman et al. (2024).

6.4.2 RNA-Based Therapeutics

Compared with DNA, RNA is a labile molecule albeit one with great conformational diversity. The high-profile development of mRNA-based vaccines during the COVID pandemic has shown that the challenge of delivering intact nucleic acids to cells can be met successfully. Other types of RNA-based medicines have preceded their use, including small interfering RNA (siRNA), antisense oligonucleotides (ASOs), aptamers and guide RNAs for the CRISPR Cas9 system. Androsavich (2024) has reviewed the RNA therapeutics field in detail, covering the introduction of novel medicines as well as the pitfalls inherent in RNA drug development. The latter include stability against nucleases, immunogenicity and manufacturing issues. Table 6.9 lists some approved RNA-based drugs (adapted from Androsavich 2024). The list gives some idea of the current activity in the field, although some of the medicines have been discontinued. In terms of market size and reach, obviously the COVID-19 vaccines have been the most successful by far.

The following sections describe the main RNA therapeutic modalities that are currently of clinical interest and have the potential for future development.

Table 6.9 Approved RNA-based drugs

Modality	Name	Target	Yr approved
mRNA	Elasomeran	COVID-19 vaccine	2019
	Tozinameran	COVID-19 vaccine	2021
saRNA	ARCT-154	COVID-19 vaccine	2023
siRNA	Vutrisiran	TTR	2022
	Inclisiran	PCSK9	2021
	Lumasiran	HAO1	2020
	Givosiran	ALAS1	2019
	Patisiran	TTR	2018
	Nedosiran	LDHA	2023
ASO	Tofersen	SOS1	2023
	Nusinersen	SMN2	2016
	Eplontersen	TTR	2023
	Volanesorsen	APOC3	2019
	Inotersen	TTR	2018
	Mipomersen	APOB	2013
	Viltolarsen	DMD exon 53	2020
	Casimersen	DMD exon 45	2021
	Golodersen	DMD exon 53	2019
	Eteplirsen	DMD exon 51	2016
Aptamer	Pegaptanib	VEGF inhibitor	2004
	Avacincaptad pegol	C5 inhibitor	2023

Adapted from Androsavich (2024) with permission

saRNA self-amplifying RNA, *siRNA* small interfering RNA, *ASO* antisense oligonucleotide

6.4.2.1 mRNA Therapeutics

As mentioned earlier, mRNA vaccines against the SARS-Cov-2 virus have provided a convincing proof of concept that mRNA delivery can be successfully achieved. Apart from further vaccine design for other infectious diseases or cancer, mRNA delivery is showing potential for treating chronic conditions such as syndromes caused by deficiencies in key metabolic enzymes. For example, Koeberl et al. (2024) have conducted a phase I/2 trial of mRNA-3927, a dual mRNA therapy for delivery of propionyl-coenzyme A carboxylase α or β (PCCA) and (PCCB) to the liver in patients with the life-threatening *propionic acidaemia* disorder. The disorder is characterised by metabolic decompensation events (MDEs) that often present in neonates with vomiting, dehydration, weight loss, feeding difficulties and acute deterioration in the individual's general clinical condition. Although the study was not complete at the time of publication, increases in PCCA and PCCB mRNA and decreases in MDEs were observed, suggesting some clinical benefit in this early-stage trial.

Although mRNA therapeutics are straightforward to produce in principle, in practice some important design parameters must be considered as described in detail by Metkar et al. (2024). These include:

Codon choice

Due to the redundancy of the genetic code, the number of codons that could be used for mRNA design is astronomical. For example, the 1273-amino acid SARS-CoV-2. spike protein could use 10^{632} possible sequences. Luckily codon bias can be analysed statistically to select sequences likely to have optimal translation efficiency.

RNA folding

The nucleotide sequence can determine the way in which the mRNA folds which has a number of consequences, including translation efficiency, mRNA splicing and even storage.[10]

Reducing immunogenicity

As part of the innate immune system, pattern recognition receptors (PRRs) have evolved to recognise foreign RNA and trigger inflammation. Toll-like receptors 7 and 8 (TLR7/8) are activated by single-stranded uridine, which is problematic for mRNA therapeutics, so the nucleotide is often substituted with analogues, the most common being 1-methyl-pseudouridine (m1Ψ). This substitution effectively silences the immune response to mRNA, but the resulting changes in 3-D structure compared with the native molecule have to be considered.

---

[10]Important for reducing the need for complex cold chain storage as was the case for COVID19 vaccines.

5′UTR design

Eukaryotic translation initiation mostly occurs through binding of a preinitiation complex to a 7-methyl-G-5′ppp5′-N cap then scans for the initiation codon (normally AUG). The structure of the 5′ untranslated region (5′UTR) determines the rate of translation; Unstructured 5′UTRs generally favour higher translation initiation rates which can be countered by increased mRNA degradation, so the 5′UTR sequence is an important consideration when designing the therapeutic molecule.

Leaky scanning

Out of frame or downstream recognition of AUG start codons (so called *leaky scanning*) is common in endogenous mRNAs and must be considered with synthetic constructs. Sequence context is important for correct initiation, the optimal being the Kozak consensus sequence GCCRCCAUGG.

3′UTR design

The 3′ untranslated region (3′UTR) can contain regulatory elements controlling subcellular localization, translation efficiency and mRNA stability, all of which need to be considered for therapeutic mRNA design. The 3′UTR affects translation initiation and mRNA stability as in long mRNAs; intramolecular base pairing juxtaposes the 5′ and 3′ ends allowing poly(A) binding protein at the 3′ end to interact with the eIF4F factor at the 5′ end. Other 3′ elements are also used for enhancing translation, and some can delay mRNA degradation. For example, a ~ 20- to 23-nucleotide post-transcriptional regulatory element (PRE) increases mRNA half-life by poly(A) tail extension. The strong immune response generated with Moderna's mRNA-1273 vaccine against SARS-CoV-2 was due in part to increased antigen production related to a poly(A) tail extension driven by the TENT5 enzyme (Metkar et al. 2024).

mRNA half-life extension

A number of ways are available for extending the half-life of mRNAs (and therefore the amount of protein produced *in vivo*). These include acting directly on the mRNA degradation machinery to create circular RNAs (circRNAs) that are resistant to exonucleolytic (but not endonucleolytic) decay. Such RNAs have extended half-lives, but not enough to make a substantial difference to mRNA therapies. However, if the mRNA is self-amplifying through co-delivery with viral replication machinery (i.e. *self-amplifying RNA* (saRNA) it should be possible to use lower mRNA doses and create a longer- lasting effect. This is a promising approach, but one with safety concerns due to the viral components involved.

Table 6.10 RNA editing companies

Name	Company	Modality	Phase
WVE-006	Wave Life Sciences/ GSK	SERPINA1/AAT mRNA editor	Phase I
AX-1412	ProQR	B4GALT1 mRNA editor	Late 2024/early 2025
AX-0810	ProQR	NTCP mRNA editor	Late 2024/early 2025
KRRO-110	Korro Bio	SERPINA1/AAT mRNA editor	IND in 2024
NA	ADARx	SERPINA1/AAT mRNA editor	Preclinical
NA	AIRNA	SERPINA1/AAT mRNA editor	Preclinical
Rett Syndrome	Vico Therapeutics	MECP2-R255X mRNA editor	Preclinical

Adapted from Mullard (2024d) with permission

6.4.2.2 RNA Editing

ADARs

DNA editing acts directly on the genome to change nucleotide sequences but changing mRNA sequences by RNA editing has the potential to be safer through more control over the dosing frequency. The relevant enzymes are adenosine deaminases acting on RNA (ADARs) and mentioned briefly in Chap 5. These editors bind mRNA and convert adenosine (A) into inosine (I) which is then read as a G by the cell's translation machinery. Endogenous ADARs act on double-stranded RNA formed by long RNAs folding back on themselves. This provides the opportunity to introduce oligonucleotides to target ADARs to specified positions on mRNAs of choice (Pfeiffer and Stafforst 2023). Some companies with RNA editing programmes based on this technology are listed in Table 6.10.

Pre-mRNA *trans*-splicing

Unlike the small base changes made in mRNAs using ADARs, RNA *trans*-splicing using the spliceosome can act over several kilobases, and using mRNA, can avoid the need to package DNA in limited capacity AAV vectors (Doi et al. 2024). The *trans*- splicing process is illustrated schematically in Fig. 6.14 (adapted from Mullard 2024e).

As with single base RNA editing programmes, there is some early clinical activity in this area. Ascidian Therapeutics are developing a *trans*-splicing product ACDN-01 designed to correct mutations in the ABCA4 transporter responsible for Stargardt disease, a form of macular degeneration. Other gene therapy approaches are being tried for this disease, but the *trans*-splicing technology means that instead of having

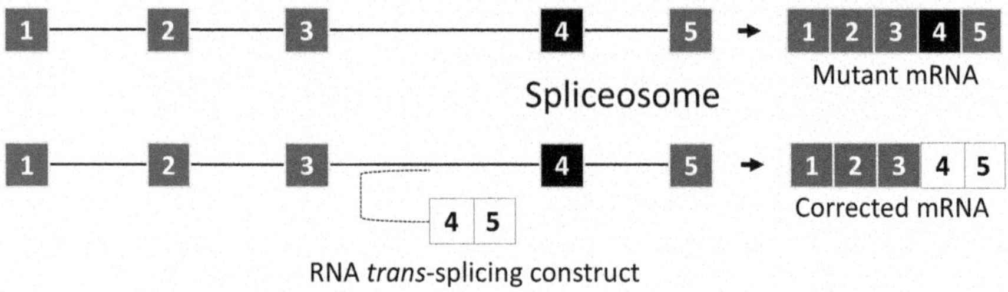

Fig. 6.14 Correction of mutated exon (4) using trans-splicing construct

Table 6.11 Approved siRNA therapeutics

Name	Target	Chemistry
Patisiran	Transthyretin	Partially modified siRNA
Givosiran	Delta ALA	ESC siRNA-GalNAc conjugate
Lumasiran	Glycolate oxidase	ESC siRNA-GalNAc conjugate
Vutrisiran	Transthyretin	ESC siRNA-GalNAc conjugate
Inclisiran	PCSK9	ESC siRNA-GalNAc conjugate
Nedosiran	LDHA	Dicer-substrate siRNA-GalNAc conjugate

From Jadhav et al. (2024)
ESC enhanced stabilisation chemistry

to load a full length 6.8Kb ABCA4 gene into AAV,[11] a shorter cDNA containing the first 22 exons can be used to edit the mRNA directly.

6.4.2.3 RNA Interference

Small Interfering RNA (siRNA)

RNA interference, its mechanism of action and application to drug target discovery has been discussed in Chap. 5. The problems surrounding the use of small interfering RNA (siRNA) inhibitors as therapeutics are the same as with other nucleic acid-based medicines, i.e. RNA stability, immunogenicity and organ-specific delivery. Unmodified siRNAs are almost immediately degraded *in vivo* (>50% within a minute), and double-stranded siRNAs can trigger innate immune responses (Tang and Khvorova 2024). Although the development of siRNAs has taken many years because of the issues mentioned above, there are now six approved siRNA drugs in the clinic (Table 6.11). Patisiran was introduced as a partially modified siRNA for the treatment of transthyretin amyloidosis, but its dosing was compromised by 5′-exonuclease activity in the endo-lysosomal compartment. Enhanced stabilisation chemistry (ESC) faces the challenge of reducing degradation while retaining the biological activity of the siRNA molecule. Alnylam, the pioneers in the field of siRNA therapeutics, have used 2′ ribose modifications (2′-fluoro/2′-OMe patterns) in later molecules, along with other modifications described by Jadhav et al. (2024).

[11] Which can only be achieved through using separate AAV constructs.

The result is a dramatic dose reduction for siRNAs with the same sequence; where revusiran required 28 grams annual dose compared with the 100 mg annual dose required by vutrisiran.

Two further points arise from the molecules listed in Table 6.11: GalNac is a ligand for the asialoglycoprotein receptor (ASGPR) which is highly expressed on hepatocytes and can be conjugated to siRNAs for targeting to the liver. Secondly, Inclisiran targets PCSK9 to lower LDL-cholesterol and therefore represents a move away from rare diseases to address a much larger population.

The expansion of siRNA therapies into other disease areas will require targeting to organs in addition to the liver. Ongoing programmes include antibody-siRNA conjugates for the transferrin receptor on muscles, Arrowhead Pharmaceuticals' integrin-siRNA conjugates for lung delivery, and Centyrins for delivery of siRNAs to multiple tissues (Klein et al. 2021).

Finally, the promising results achieved with siRNA therapeutics makes them good candidates for combination therapies in difficult-to-treat conditions such as cancer.

Antisense Oligonucleotides (ASOs)

Antisense oligonucleotides are short single or double stranded molecules designed to interact with mRNA via Watson-Crick base pairing and influence the resulting translated protein through direct inhibition of mRNA or modification of pre-mRNA splicing.[12] ASOs fall into the categories of small interfering RNAs discussed above, gapmer ASOs and splice switching ASOs (ssASOs) (Lauffer et al. 2024).

Gapmers consist of a single stranded DNA core flanked by RNA-based structures resistant to RNase H. Sequence-specific binding creates a DNA:RNA hybrid of the gapmer core with the target mRNA transcript being degraded by RNAse H which recognises DNA:RNA hybrids. This is illustrated in Fig. 6.15 along with part of the structure of the gapmer Tofersen, an ASO that targets misfunctional SOD1 to treat amyotrophic lateral sclerosis. Tofersen has twenty bases configured as RNA-DNA-RNA (5–10-5) with a mixed backbone to resist degradation that includes 2'-O-(2-methoxyethyl)-D-ribose and phosphorothioate substitutions (both arrowed).

Splice switching ASOs (ssASOs) have shown how restoring protein function can be achieved by targeting pre-mRNA splicing. For example, the ssASO Nusinersen increases SMN protein in patents with spinal muscular atrophy and Casimersen ASO skips mutations in exon 45 of the dystrophin gene leading to restoration of functional protein in muscular dystrophy patients.

[12] See Table 6.9 for examples.

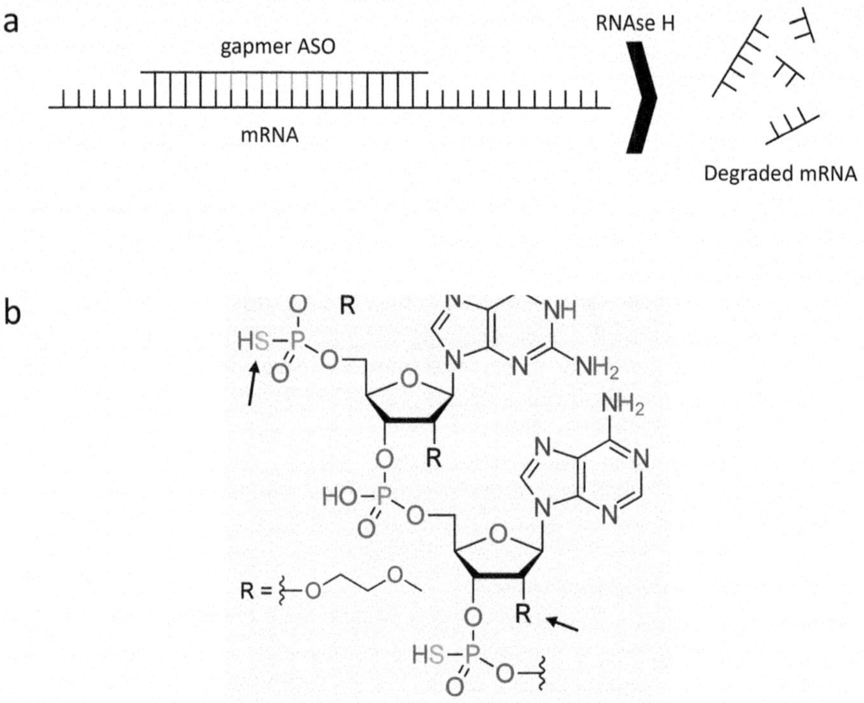

Fig. 6.15 (**a**) Binding of gapmer ASO to mRNA creates DNA-RNA hybrid susceptible to degradation by RNAse H. (**b**) part of structure of 20 nt gapmer Tofersen showing modifications to reduce action of nucleases

6.4.2.4 Aptamers

Aptamers (Latin: *aptus*, fit; Greek: *meros*, part or region) are single-stranded DNA or RNA molecules that bind directly to protein targets rather than regulate their production at the level of genes or mRNA (reviewed by Nimjee et al. 2017). Aptamers are selected from large libraries using *systematic evolution of ligands by exponential enrichment* (SELEX) yielding candidates with nanomolar to picomolar affinities. As with other nucleic acid-based drugs, nuclease degradation can be reduced by modifications to the nucleoside components and their half-lives increased using polyethylene glycol (PEGylation). Antidote oligonucleotides can be designed to suppress any adverse events that might arise in patients treated with a therapeutic aptamer candidate. Despite these attractive characteristics, only one aptamer, has been marketed as a drug; Pegaptanib was developed as an aptamer against vascular endothelial growth factor (VEGF) to treat macular degeneration and approved in 2004, but largely superseded by antibody therapies. However, intraocular therapies using aptamers are still considered viable, with Iveric Bio (now part of Astellas) receiving approval for avacincaptad pegol a complement C5 inhibitor for treating geographic atrophy (GA) secondary to age-related macular degeneration.

Basking Biosciences are developing a novel RNA aptamer acting against von Willebrand factor (vWF) for use as an antithrombotic. It is possible that more aptamer drugs will appear over time, but they must compete with other modalities such as antibodies and engineered molecules such as macrocycles etc. In any event, the relative ease by which high affinity specifically targeted aptamers can be created makes them useful for diagnostic applications.

6.4.2.5 RNA Delivery

Drugs based on RNA chemistry could never have been viable without a means to deliver them to target cells in an undegraded state. The development of liposomes in the last part of the twentieth century meant that mRNA could be protected by a lipid coat prior to fusion with cell membranes and delivery into the cytoplasm, a system later optimised by using cationic and ionizable lipids. The development of lipid nanoparticles (LNPs) for mRNA delivery has been reviewed in detail by Hou et al. (2021). There is a large choice of lipid components and modifications (e.g. with polyethylene glycol) as well as hybrid nanoparticles that incorporate polymers such as polyethyleneimine, protamine and polyaspartamide derivatives. Exosomes and cell membranes can also be used to deliver mRNA. However, these materials can provoke adverse immune reactions and alter pharmacokinetic and pharmacodynamic profiles. Targeting to specific organs can be achieved by selecting an optimal lipid structure as illustrated by Swingle et al. (2025) for delivering VEGF mRNA to the placenta in a mouse model of pre-eclampsia. *In vivo* screening for lipids that achieved placental rather than hepatic delivery of LNPs was conducted using DNA barcoding of large LNP libraries to evaluate LNPs in more complex biological environments (see Dahlman 2017). From a library of 98 LNP formulations, one gave an order of magnitude greater placental delivery than clinically used LNPs for other applications.

Other organ targeting LNPs are being developed by ReCode Therapeutics using Selective ORgan Targeting (SORT) LNPs to deliver RNA constructs to the lung and other organs. The company is conducting clinical trials with RCT1100, an inhaled mRNA-LNP for Primary Ciliary Dyskinesia caused by pathogenic mutations in DNAI1, a gene encoding a protein essential for ciliary movement.

6.4.2.6 Manufacturing

The growing success of RNA-based medicines and the prospect of market expansion from rare to common diseases has implications for large-scale production of nucleic acids and LNP components.

mRNA-LNPs

Therapeutic mRNAs are produced by *in vitro* transcription from a linearised DNA template which is later removed by DNAse digestion. The mRNA is capped and purified chromatographically or using microbeads to remove reagents and any double-stranded molecules that could provoke an immune response. LNP-mRNA formulations are commonly manufactured by rapidly mixing ethanol phase (lipids) and aqueous phase (mRNA) at defined pH and flow rates (Hou et al. 2021). The formulations are purified to remove solvents etc., followed by filtration and freezing.

One of the main issues with the mRNA vaccines for COVID-19 was the need to store them at low temperatures. This increases the complexity of the cold chain, particularly for mass vaccination, so it is important that formulations are developed to allow storage at higher temperatures.

The section above is concerned with mRNA production, but oligonucleotide-based drugs have their own manufacturing requirements. According to Obexer et al. (2024) solid phase phosphoramidite chemistry is used to extend nucleotide sequences using a series of iterative coupling, capping, oxidation, and deprotection steps. Solid supports restrict the synthesis to <10 Kg batches while using large volumes of acetonitrile (~1000 Kg per Kg of oligonucleotide). As a result, currently marketed RNA-based drugs are typically administered with ~90% purity and, in the case of phosphorothioate-modified sequences, as complex mixtures of stereoisomers. However, considering that synthesis of a 20mer for example, comprises ~80 chemical steps this is an impressive achievement. Phosphoramidite chemistry is being scaled up using liquid-phase strategies and chromatography in the aqueous phase thereby reducing the need for acetonitrile. Obexer et al. (2024) discuss newer synthetic approaches, including biocatalysis, that in addition to providing sustainable oligonucleotide medicines in industrial quantities, could also enable the direct synthesis of longer mRNA products with total flexibility over the chemistry of individual nucleotides.

6.5 Cell Therapies

The idea of using live cells as therapeutics has been discussed for many years and was put into practice in the mid twentieth century in the form of bone marrow transplantation. However, with a biopharmaceutical industry almost exclusively focussed on small molecule drugs, new approaches such as therapeutic proteins and gene therapies took a long time to be accepted as mainstream products. This century has seen a big change with the inclusion of cell therapies in major company pipelines, largely because they work in situations where thus far all other options have failed. They also provide powerful delivery vehicles for gene therapies, both as stem cells and even microbes for drug targeting. Advances in bioelectronics means that cells could be programmed to deliver therapeutics thereby adding an extra level of sophistication (see Rivnay et al. 2024).

Many of the issues with gene therapy in terms of safety, cost and market size are also present with cell therapy, with the extra challenge of having to engineer patients' cells *ex vivo* in dedicated facilities. The future of these therapies will depend upon whether these challenges can be overcome and whether alternative modalities such as antibodies with similar efficacy will be developed instead.

Table 6.12 shows a list of relevant FDA-approved cell therapies by category (taken from FDA Cell and Gene Therapy Approvals 2024).

Table 6.12 FDA-approved cell therapies (as of December 2024)

CAR T-Cells	Engineered stem cells	Autologous T-Cells	Islet Cells	Autologous stromal cells
Abecma	Casgevy	Amtagvi	Lantidra	Ryoncil
Aucatzyl	Lenmeldy	Provenge		
Breyanzi	Lyfgenia	Tecelra		
Carvykti	Skysona			
Kymriah	Zynteglo			
Tecartus				
Yescarta				

The following sections cover the types of cell therapies listed in the Table and the ways in which they are being developed for use against hard-to-treat diseases.

6.5.1 Stem Cells

6.5.1.1 Haematopoietic Stem Cells

Haematopoiesis is the differentiation of haematopoietic stem/progenitor cells (HSPCs) into the different blood cell lineages leading ultimately to lymphoid and myeloid cells, the latter including red blood cells through the erythroid lineage. The success of bone marrow transplantation over many decades bears testimony to the therapeutic benefit of replacing diseased blood cells (e.g. in leukaemia) with normal cells from a suitable donor (allogeneic transplantation). The ability to introduce or remove genes from HSPCs forms the basis of engineered autologous stem cell therapy, (i.e. using cells taken from the patient, examples of which were discussed earlier in Sect. 6.4.1. and are given in Table 6.12). In the long term, it will be important to know whether the clinical benefits are durable and how the replicative stress imposed by haematopoietic reconstitution might affect the HSPCs' multi-lineage potential, dynamics of lineage commitment, long-term safety, and variation in different disease conditions. For example, sickle cell disease patients have an increased frequency of haematopoietic clones bearing driver mutations associated with myeloid cancer or clonal haematopoiesis, as observed in untreated and treated patients receiving gene therapy (Calabria et al. 2024). In an attempt to address these questions, these authors have tracked the fate of genetically modified HSPCs in 53 patients with three different diseases up to 8 years post transplantation. Lentiviruses were used to correct β-thalassaemia (β-globin expression in erythroid cells), metachromatic leukodystrophy (MLD) (arylsulfatase A in myeloid cells), and Wiskott-Aldrich syndrome (WAS) (WASp expression in lymphoid cells). They found that the long-term output, lineage commitment of HSPCs, and accumulation of somatic mutations, were strongly modulated by the patients' genetic backgrounds, possibly to compensate for the demands posed by the specific clinical condition. Once genetically modified HSPCs had reached stability, they contributed to about 50% of the haematopoiesis for all conditions while the remaining clones showed disease-specific commitment. Haematopoiesis is maintained by many lineage-committed HSPCs long after transplantation and can persist for several years with no evidence of clonal dominance. These results are encouraging when considering future gene therapy, either using HSPCs modified by lentiviruses or by other means (e.g. lipid nanoparticles).

6.5.1.2 Tissue-Specific Stem Cells

The foundations of regenerative cell therapy were established in 1998 with the generation of human embryonic stem cells (hESCs) which could be differentiated into different cell types for therapeutic use, such as pancreatic beta cells, retinal pigment epithelial cells, cardiomyocytes, and neurons. The development of human induced pluripotent stem cells (hiPSCs) from somatic cells in 2007 made it possible to create autologous cells for implantation that avoided the problem of immune rejection. Despite the great clinical promise for treating diseases such as Type I diabetes and Parkinson's Disease, progress has been slow for reasons outlined in a detailed review of the field by Kirkeby et al. (2025). Initial trials employed low numbers of cells and their administration to immune privileged sites such as the CNS and eye. Geron Corporation was the first to attempt clinical development of a cell product (oligodendrocyte progenitor cells) derived from hESCs, filing a 22,000-page IND to the regulators. After having to conduct safety evaluations for 2 years, followed by a one-year clinical trial, the project was terminated for financial reasons. Since that unpromising start, as of December 2024, Kirkeby et al. (2025) have counted 83 hPSC-derived products undergoing testing in 115 clinical trials world-wide with at least 1200 patients and 34 different indications. These data demonstrate that for properly regulated stem cell trials the hPSC-derived products are safe and well tolerated. Reassurance on this point, plus overcoming GMP issues will determine the level of interest in stem cell therapies by investigators in companies and academic institutions. Some indications that this may be the case are given in the case of Islet cell replacement in patients with Type I diabetes where the FDA has approved CellTrans Inc's cadaver-derived Islet cell preparation Lantidra. This success has prompted the use of stem cells to produce Islets rather than relying upon cadaver tissues; promising results have been obtained from a Peking University study using Islets formed *in vitro* from chemically induced pluripotent stem cells and transplanted into abdominal muscles (Mallapaty 2024a). Vertex Inc. developed a similar product VX-880 through its acquisition of the originator Semma Therapeutics; the HSC-derived mature islets were delivered directly through the hepatic portal vein in immunosuppressed patients almost all of whom had reduced, or no requirement for exogenous insulin. In late 2024, the company announced plans for a pivotal phase I/II/III trial highlighting the fact that a mature biopharmaceutical company has committed to marketing a complex cell therapy of this type. Patients receiving transplanted cells are normally exposed to immunosuppressive regimes, but Vertex's VX-264 cells may avoid this through being encapsulated in a protective *channel array device* designed to protect cells from the immune system while enabling growth of blood vessels in membrane-covered channels.

Diabetes is only one of many disease areas being investigated for applications of HSCs as reviewed by Kirkeby et al. (2025). The need for alternatives to ES cells has focused attention on iPSCs that can be differentiated with chemical treatment avoiding the issues raised using viral transfection. Another potential safety issue relates to changes in the genome and epigenome of induced cells compared with their normal counterparts; with years of experience built up so far, it would appear that this is unlikely to be a significant problem as cells are screened extensively before administration to patients (Abbott 2025).

6.5.2 Cells of the Immune System

Antibodies and other protein effectors of *humoral immunity* are well established therapeutics; *cellular immunity* in the form of T lymphocytes, NK cells, monocytes etc. is highly relevant to cancer therapy given the role of these cells in immune surveillance; development of cell therapies based on this system is one of the hot areas in drug development, with Saez-Ibañez et al. (2024) reviewing the state of clinical development. As with other cell therapies, the products can be taken unmodified from donors or enhanced using genetic engineering to tailor the immune response to the target of interest.

6.5.2.1 T-Cells

Unmodified Cells

Adoptive cell transfer (ACT) cancer immunotherapy is achieved by isolating and expanding tumour-infiltrating lymphocytes (TILs) from a patient and reinfusing them to target the tumour cells. A pioneer in this aspect of immunotherapy (Rosenberg 2024) has described recent progress in the area decades after the first clinical trials using this approach:

The major challenge confronting the future development of ACT immunotherapy using TILs is the development of effective therapies for patients with metastatic solid epithelial cancers that cannot be cured by any available treatment and that account for 90% of all deaths from cancer. Current systemic drug treatments often substantially prolong life for patients with many types of metastatic epithelial cancers, but cures are exceedingly rare. The exquisite specificity and sensitivity of the immune system has the potential to control the last cancer cell, as illustrated by the many patients with metastatic melanoma alive and disease free more than 20 years after treatment with Interleukin-2 (IL-2) or TILs.

IL-2 is used to expand T-cell numbers prior to infusion into patients (the problem of IL-2 toxicity has been discussed in Sect. 6.3.1.1). The first cellular therapy for a solid tumour, Lifileucel, was developed by the US National Cancer Institute for melanoma and approved by the FDA in early 2024. While showing promise for cancer treatment, the highly individualised therapies are not amenable to scaling in the same way as other biopharmaceutical products. Like gene therapy, the logistical issues in development and manufacturing, as well as pricing, will determine how cell therapies of this type will influence clinical practice in the future.

Engineered Natural T-Cell Receptors (TCRs)

The affinity of T-cell receptors for their target antigen (presented as peptides on MHC molecules) is modest compared with antibodies; this is a problem for T-cell recognition of often weakly immunogenic tumour cell antigens.[13] One way round this problem is to engineer TCRs with enhanced affinities into autologous T-cells to boost tumour recognition. Adaptimmune has gained accelerated FDA approval for the use in synovial sarcoma of autologous patient T-cells engineered to enhance TCR affinity for the MAGE4 tumour antigen (afamitresgene autoleucel).

Another approach has been taken by Immunocore Ltd. whose ImmTac proteins are engineered TCRs with high affinity for the cancer target linked to a single chain antiCD3 thus binding the constructs to the patient's circulating T-cells (Lowe et al. 2019). KIMMTRAK (Tebentafusp) is an ImmTac product which has been approved for treating uveal melanoma (Nathan et al. 2021).

Gamma Delta (γδ) T-Cells

γδ T-cells have broader reactivities than the predominant T-cell subset bearing TCRs with αβ chains. Importantly, γδ cells are not MHC restricted and therefore can be generated and used off-the-shelf from healthy donors. The current state of progress in developing immunotherapies based on γδ T-cells is reviewed by Hayday et al. (2024). In addition to adoptive cell transfer of unmodified cells, γδ TCR genes can be engineered into bispecific antibodies or chimaeric antigen receptor (CAR) cells with γδ TCRs to potentially improve efficacy. Company activity in the immunotherapy area includes In8Bio whose γδ T-cells processed with their Deltex technology are in clinical trials for acute myeloid leukaemia (AML).

Chimaeric Antigen Receptor (CAR) T-Cells

Of the different FDA-approved cell therapies listed in Table 6.12, CAR T-cells make up the largest group. This is because these genetically engineered autologous patient-derived cells have been shown to be curative in 50% of leukaemia, lymphoma and melanoma cases treated. The first CAR therapy Kymriah, was developed by Novartis and approved by the FDA in 2017 for treating B-cell lymphomas and B-cell acute lymphoblastic leukaemia. The CAR strategy is based on boosting T-cell recognition of tumours by engineering a receptor construct that binds intact tumour cell antigens without MHC involvement thus bypassing native T-cell receptors. The target antigen has to be carefully selected to avoid off target effects, which is why the CD19 antigen was chosen as it is B-cell restricted and unlikely to cause general toxicity. The CAR construct is shown in Fig. 6.16 showing the separate components of normal T-cell recognition and signalling that are engineered into a compact module. Other CAR therapies have been approved since Kymriah, and clinical experience is being acquired through treating thousands of patients, mostly for B-cell cancers. Much work is underway to expand the range of these potentially

---

[13] Cold tumours caused by weak TCR affinity and an immunosuppressive environment.

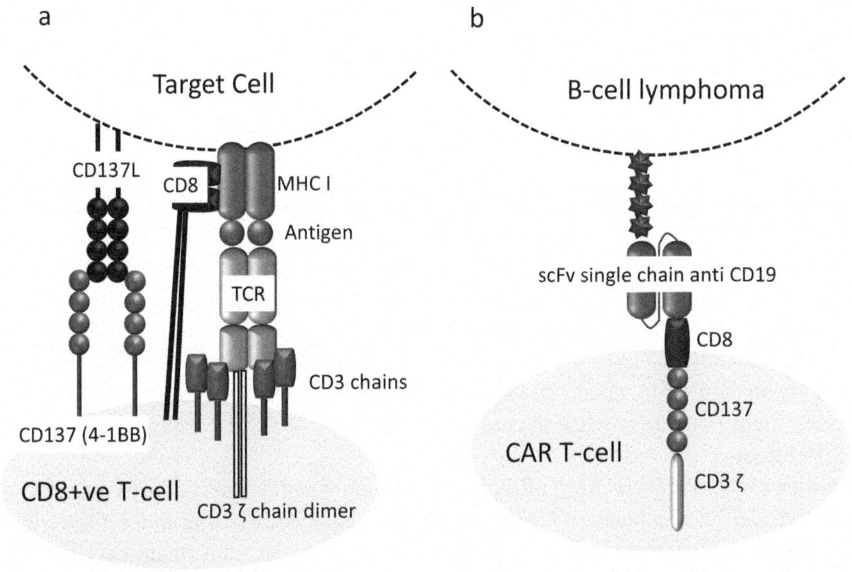

Fig. 6.16 Engineering of CAR T-cells. (**a**) Interactions between CD8+ve cytotoxic T-cell and peptide antigen presented by MHC I molecules on a target cell. Antigen recognition by the T-cell receptor (TCR) leads to cell activation via the CD3 complex and dimerisation of the CD3 ζ chain. The co-stimulatory interaction between CD137L and CD137 is required for T cell activation and the CD8 interaction with MHC I amplifies the response. (**b**) The features used in (**a**) are engineered into autologous CD8+ve T-cells except that the target is an unprocessed surface antigen (in this case CD19) on the surface of a B-cell lymphoma. Antigen recognition is via an engineered antiCD19 antibody domain allowing direct binding. The signalling and costimulatory molecules are all packaged into a single construct

curative products to include other diseases including cancers with solid tumours. These include CAR T-cells against CD19 for the autoimmune diseases systemic lupus erythematosus (SLE), idiopathic inflammatory myositis, and systemic sclerosis (Müller et al. 2024) and the EGFRvIII protein restricted to glioblastoma cells (Choi et al. 2024).

Despite the clinical success of CAR T-cells there are significant problems to be overcome if they are to become mainstream therapies as outlined below:

Increasing the repertoire of target antigens

The ability to engineer different functions into CAR T-cells allows for flexibility in target identification. As summarised by Davila and Brentjens (2024), in mouse models, CAR T-cells can be targeted to cross the blood-brain barrier by detecting the brain-specific brevican (BCAN) protein using an engineered sensor domain of synNotch receptors in a CAR construct. Once in the brain, the CAR recognises glioma-associated ephrin type A receptor 2 (EphA2) and interleukin-13 receptor α2 (IL13Rα2) specific to glioblastoma cells. The CNS-targeted cells efficiently cleared primary tumours and persisted in the brain weeks to months after tumour regression.

Adverse Drug Reactions

Cytokine release syndrome can be a fatal consequence of CAR-cell infusion and is closely monitored during therapy. The standard of care involves corticosteroids or Tocilizumab, a neutralizing antibody for interleukin-6 receptor α (IL-6Rα), the latter because highly elevated levels of IL-6 have been observed in CAR T-cell patients. Again, using the flexibility of genetic engineering T-cells, one approach is to programme human CAR T-cells directed against CD19 to release IL-6 binding proteins in situ (Yin et al. 2024). As well as working in a humanised mouse model, the procedure actually increased the efficacy of CAR treatment against the CD19 target.
Another concern that has been raised in the prevalence of any secondary cancers that may have arisen as a result of CAR therapy. This is an issue that is under discussion, but with no clear answer at present (Willyard 2024). As with treatments for any serious disease, it is a question of weighing up the risk/benefit ratio.

Clonal exhaustion of CAR T-cells

T-cell exhaustion results from their differentiation into a defined cell subset with impaired proliferation and survival after sustained activity against a target. This occurs during CAR T-cell treatment, particularly with solid tumours, so attempts are being made to understand and inhibit the exhaustion mechanism to allow sustained T-cell activity. Memory-associated genes in CAR T-cells are linked to their long-term persistence in patients and clinical efficacy, with the FOXO1 transcription factor responsible for promoting memory and restraining exhaustion in human CAR T cells (Doan et al. 2024). CAR T-cells could therefore be directly engineered to overexpress FOXO1 and prevent exhaustion, something which could also occur through inducing IL-15 expression, a cytokine known to upregulate FOXO1 (Chan et al. 2024).

Manufacturing issues and associated costs

CAR T-cells are produced autologously from donor patients, which is not ideal when considering the conventional off-the-shelf nature of pharmaceutical products. Allogeneic CAR T-cells are being developed for this purpose but are yet to make a significant impact because of safety issues, but this may change in the future (see Moradi et al. (2023) for a review, including descriptions of CAR T-cell manufacturing). These manufacturing costs, and associated medical support, are considerable in western markets and generally beyond the reach of lower income countries. For example, to quote from Mallapaty (2024b) the price for sufficient lentivirus for a trial of 50 people in the USA can be up to $800,000. Given the incentive, however it is possible to make significant cost reductions. The Indian NexCAR19 product targets CD19 like others but can be produced for $30,000–$40,000 compared with $370,000 to $530,000 with existing commercial products. The cost reduction of this magnitude is promising for a future where CAR-based therapies will hopefully become available for a broader range of diseases than is currently the case.

In Vivo CAR T-Cells

The problem of having to process cells *ex vivo* would be eliminated if CAR-T cells could be produced directly in patients. The approach is essentially one of converting cell therapy with modified T-cells into gene therapy where the modification occurs with reprogramming cassettes. Patients are not immunosuppressed, unlike the case with CAR T-cells administered to patients treated with chemotherapy to remove endogenous immune cells; in addition, this may enhance the efficacy of the *in vivo* CAR treatment due to the activity of an intact immune system. The relevant gene cassettes can be delivered using lentiviruses or LNPs with mRNAs. Some companies and their *in vivo* CAR therapies under development are listed in Table 6.13 (see also Mullard 2024f). Most of the CAR constructs are directed towards B-cells as with current *ex vivo* products The myeloid CAR cells listed, from Myeloid Therapeutics and Carisma/Moderna, will be discussed later.

The molecules used for targeting T-cells are shown in brackets, with CD3 the most common, but also CD58 and CD80 used by Umoja Biosciences to enhance co-stimulation. Umoja also added an artificial rapamycin-activated cytokine receptor (RACR) system that releases IL-15 upon administration of the immunosuppressive drug rapamycin. As mentioned earlier, IL-15 upregulates FOXO1 to overcome T-cell exhaustion, thereby extending the lifespan of the active CAR cells. For

Table 6.13 *In vivo* CAR therapies under development

Company	Vector	Targeting	CAR	Indication
Interius BioTherapeutics	Lentivirus	T-cells (CD7)	CD20	B cell cancers
Interius BioTherapeutics	Lentivirus	T-cells (CD7)	CD19	Autoimmune
Umoja Biopharma	Lentivirus	T-cells (CD3/58/80)	CD19	B cell cancers
Umoja Biopharma	Lentivirus	T-cells (CD3/58/80)	CD22	B cell cancers
Umoja Biopharma	Lentivirus	T-cells (CD3/58/80)	CD19	B cell cancers
Kelonia Therapeutics	Lentivirus	T-cells (CD3)	BCMA	Multiple myeloma
Capstantx	LNP	T-cells (CD8)	CD19	Autoimmune
Orbital Therapeutics	LNP	None	CD19	Autoimmune
Orna	LNP	None	CD19	B cell cancers
Orna	LNP	None	CD19	Autoimmune
Orna	LNP	None	BCMA	Multiple myeloma
Myeloid Therapeutics	LNP	Myeloid cells	TROP2	Epithelial tumours
Myeloid Therapeutics	LNP	Myeloid cells	GPC3	Liver cancer
Carisma/Moderna	LNP	Myeloid cells	GPC3	Liver cancer

targeting to T-cells using LNPs, Capstantx conjugated the latter with an antibody to CD8 for mRNA delivery. Orbital Therapeutics and Orna used circular RNA to boost production of their CAR proteins.

6.5.2.2 Myeloid and NK Cells

Myeloid CARs

Myeloid Therapeutics and Carisma Therapeutics (partnered with Moderna) are developing *in vivo* CAR constructs in myeloid cells for cancer. Myeloid cells, unlike T-cells can readily infiltrate solid tumours leading to phagocytosis and activation of inflammatory pathways. If these approaches are successful with no major safety liabilities, myeloid cell therapeutics may make a significant contribution to cancer therapy.

Dendritic Cells and Cancer Vaccines

Dendritic cells (DCs) are considered to be *professional* antigen presenting cells being mostly localized in tissues until recruited to sites of infection. DCs are very efficient in presenting antigens on MHC class I and II molecules to activate CD8 and CD4+ve T-cells respectively. These properties, as well as the co-stimulatory signals provided by DCs, has made them attractive candidates for vaccinating against solid tumours. In practice, DCs are created *ex vivo* from leukocytes (GM-CSF-treated) and monocytes (IL-15 treated) then activated with antigens *in vitro* prior to administration to the patient (Santos and Butterfield 2018). The first DC vaccine to be approved by the FDA (in 2010) was Provenge, for prostate cancer using DCs simultaneously generated and pulsed with antigen using a fusion protein of GM-CSF and prostatic acid phosphatase. More recently, Diakonos Oncology Corp have received fast track FDA approvals for the DC vaccine DOC1021 for both pancreatic cancer and glioblastoma. Improvements in antigen loading (including with mRNA) plus the possibility of genetically modifying the DCs to enhance tumour penetration etc., suggest that DC vaccines will become more widespread in the future.

Natural Killer (NK) Cells

Natural killer (NK) cells are part of the innate immune response and act as a first line of defence prior to adaptive antibody and T-cell responses. NK cells target cells through recognition by receptors such as NKG2D, leading to cell death similar to that caused by cytotoxic T-cells. NK cells are readily obtainable from peripheral blood, cord blood and differentiated stem cells thus providing an off-the-shelf product. Fate Therapeutics have produced NK cells from iPS cells which have shown promise in treating solid tumours (Patel et al. 2022). NK cells can be engineered as CAR therapeutics (reviewed by Zhong and Liu 2024) and form the current pipeline of Fate Therapeutics. There are problems with CAR-NK cells however, as they are short lived compared with T-cells and readily inhibited by the immunosuppressive tumour microenvironment. Some of the strategies employed to overcome these obstacles are reviewed in Zhong and Liu (2024). A number of CAR-NK programmes are in clinical development, the first being a treatment for CD19 positive cancers reported by Liu e al. (2020).

References

Abbott E (1884) Flatland: a romance of many dimensions with illustrations by the author, A Square. Seeley & Co, London

Abbott A (2025) The age of stem-cell therapy. Nature 637:18–20. https://doi.org/10.1038/d41586-024-04160-0. Accessed 20 Mar 2025

Adaptimmune. https://www.adaptimmune.com/. Accessed 20 Mar 2025

ADARx. https://www.adarx.com/. Accessed 20 Mar 2025

Aera Therapeutics. https://aeratx.com/. Accessed 20 Mar 2025

AIRNA. https://airna.com/. Accessed 20 Mar 2025

Akilesh S et al (2007) Neonatal FcR expression in bone marrow-derived cells functions to protect serum IgG from catabolism. JImmunol 179:4580–4588. https://doi.org/10.4049/jimmunol.179.7.4580. Accessed 20 Mar 2025

Al Zubaydi S et al (2024) Reductive alkyl–alkyl coupling from isolable nickel–alkyl complexes. Nature 634:585–591. https://doi.org/10.1038/s41586-024-07987-9. Accessed 20 Mar 2025

Ali RSAE et al (2024) Machine learning advancements in organic synthesis: a focused exploration of artificial intelligence applications in chemistry. Artif Intell Chem 2:100049. https://doi.org/10.1016/j.aichem.2024.100049. Accessed 20 Mar 2025

Aliper A et al (2016) Deep learning applications for predicting pharmacological properties of drugs and drug repurposing using transcriptomic data. Mol Pharm 13:2524–2530. https://doi.org/10.1021/acs.molpharmaceut.6b00248. Accessed 20 Mar 2025

Alnylam Pharmaceuticals. https://www.alnylam.com/. Accessed 20 Mar 2025

Androsavich JR (2024) Frameworks for transformational breakthroughs in RNA-based medicines. Nat Rev Drug Discov 23:421–444. https://doi.org/10.1038/s41573-024-00943-2. Accessed 20 Mar 2025

Arrakis Therapeutics. https://arrakistx.com/. Accessed 20 Mar 2025

Arrowhead Pharmaceuticals. https://arrowheadpharma.com/. Accessed 20 Mar 2025

Ascidian Therapeutics. https://ascidian-tx.com/. Accessed 20 Mar 2025

Astellas. https://www.astellas.com/en/. Accessed 20 Mar 2025

Atomic AI. https://atomic.ai/. Accessed 20 Mar 2025

Autio KA et al (2020) Probody therapeutics: an emerging class of therapies designed to enhance on-target effects with reduced off-tumor toxicity for use in immuno-oncology. Clin Cancer Res 26:984–989. https://doi.org/10.1158/1078-0432.ccr-19-1457. Accessed 20 Mar 2025

Avectas. https://www.avectas.com/. Accessed 20 Mar 2025

Avilar Therapeutics. https://avilar-tx.com/. Accessed 20 Mar 2025

Axup JY et al (2012) Synthesis of site-specific antibody-drug conjugates using unnatural amino acids. Proc Natl Acad Sci USA 109:16101–16106. https://doi.org/10.1073/pnas.1211023109. Accessed 20 Mar 2025

Basking Biosciences. https://baskingbiosciences.com/. Accessed 20 Mar 2025

Benn J et al (2024) Aggregate-selective removal of pathological tau by clustering-activated degraders. Science 385:1009–1016. https://doi.org/10.1126/science.adp5186. Accessed 20 Mar 2025

Bicycle Therapeutics. https://www.bicycletherapeutics.com/. Accessed 20 Mar 2025

Biohaven. https://www.biohaven.com/. Accessed 20 Mar 2025

Bohacek RS (1996) The art and practice of structure-based drug design: a molecular modeling perspective. Med Chem Rev 16:3–50. https://doi.org/10.1002/(SICI)1098-1128(199601)16:1<3::AID-MED1>3.0.CO;2-6. Accessed 20 Mar 2025

Boldini D et al (2024) Machine learning assisted hit prioritization for high throughput screening in drug discovery. ACS Cent Sci 10:823–832. https://doi.org/10.1021/acscentsci.3c01517. Accessed 20 Mar 2025

Bolognesi ML, Cavalli A (2016) Multitarget drug discovery and polypharmacology. ChemMedChem 11:1190–1192. https://doi.org/10.1002/cmdc.201600161. Accessed 20 Mar 2025

Bremer PT et al (2023) Investigation of monoclonal antibody CSX-1004 for fentanyl overdose. Nat Commun 14:7700. https://doi.org/10.1038/s41467-023-43126-0. Accessed 20 Mar 2025
CACHE. https://cache-challenge.org/cache-news. Accessed 20 Mar 2025
Calabria A et al (2024) Long-term lineage commitment in haematopoietic stem cell gene therapy. Nature 636:162–171. https://doi.org/10.1038/s41586-024-08250-x. Accessed 20 Mar 2025
Capstantx. https://www.capstantx.com/. Accessed 20 Mar 2025
Carisma Therapeutics. https://carismatx.com/. Accessed 20 Mar 2025
CellTrans Inc. https://www.celltransinc.com/. Accessed 20 Mar 2025
Chan JD et al (2024) FOXO1 enhances CAR T cell stemness, metabolic fitness and efficacy. Nature 629:201–210. https://doi.org/10.1038/s41586-024-07242-1. Accessed 20 Mar 2025
Chen X et al (2023) Targeted degradation of extracellular secreted and membrane proteins. Trends Pharmacol Sci 44:762–775. https://doi.org/10.1016/j.tips.2023.08.013. Accessed 20 Mar 2025
Chin J (2017) Expanding and reprogramming the genetic code. Nature 550:53–60. https://doi.org/10.1038/nature24031. Accessed 20 Mar 2025
Cho H et al (2011) Optimized clinical performance of growth hormone with an expanded genetic code. Proc Natl Acad Sci USA 108:9060–9065. https://doi.org/10.1073/pnas.1100387108. Accessed 20 Mar 2025
Choi BD et al (2024) Intraventricular CARv3-TEAM-E T cells in recurrent glioblastoma. N Engl J Med 390:1290–1298. https://doi.org/10.1056/NEJMoa2314390. Accessed 20 Mar 2025
Chu AE et al (2024) Sparks of function by de novo protein design. Nat Biotechnol 42:203–215. https://doi.org/10.1038/s41587-024-02133-2. Accessed 20 Mar 2025
Codarri Deak L et al (2022) PD-1-cis IL-2R agonism yields better effectors from stem-like CD8+ T cells. Nature 610:161–172. https://doi.org/10.1038/s41586-022-05192-0. Accessed 20 Mar 2025
Code Bio. https://www.codebiotx.com/. Accessed 20 Mar 2025
Dahlman JE et al (2017) Barcoded nanoparticles for high throughput in vivo discovery of targeted therapeutics. Proc Natl Acad Sci USA 114:2060–2065. https://doi.org/10.1073/pnas.1620874114. Accessed 20 Mar 2025
Dai T et al (2024) Autonomous mobile robots for exploratory synthetic chemistry. Nature 635:890–897. https://doi.org/10.1038/s41586-024-08173-7. Accessed 20 Mar 2025
Davies IW (2019) The digitization of organic synthesis. Nature 570:175–181. https://doi.org/10.1038/s41586-019-1288-y. Accessed 20 Mar 2025
Davila ML, Brentjens R (2024) Synthetic gene circuits drive disease-fighting T cells. Science 386:1094–1095. https://doi.org/10.1126/science.adt9921. Accessed 20 Mar 2025
Di Martino RMC (2023) Deuterium in drug discovery: progress, opportunities and challenges. Nat Rev Drug Discov 22:562–584. https://doi.org/10.1038/s41573-023-00703-8. Accessed 20 Mar 2025
Diakonos Oncology Corp. https://www.diakonosoncology.com/. Accessed 20 Mar 2025
Ding W et al (2024) Rare codon recoding for efficient noncanonical amino acid incorporation in mammalian cells. Science 384:1134–1142. https://doi.org/10.1126/science.adm8143. Accessed 20 Mar 2025
Doan AE et al (2024) FOXO1 is a master regulator of memory programming in CAR T cells. Nature 629:211–218. https://doi.org/10.1038/s41586-024-07300-8. Accessed 20 Mar 2025
Doi A et al (2024) RNA exon editing: splicing the way to treat human diseases. Molecular therapy. Nucleic Acids 35:102311. https://doi.org/10.1016/j.omtn.2024.102311. Accessed 20 Mar 2025
Dongol Y et al (2019) Spider knottin pharmacology at voltage-gated sodium channels and their potential to modulate pain pathways. Toxins 11:626. https://doi.org/10.3390/toxins11110626. Accessed 20 Mar 2025
Dumontet C et al (2023) Antibody–drug conjugates come of age in oncology. Nat Rev Drug Discov 22:641–661. https://doi.org/10.1038/s41573-023-00709-2. Accessed 20 Mar 2025
Emert-Sedlak LA et al (2022) Antiretroviral drug discovery targeting the HIV-1 Nef virulence factor. Viruses 14. https://doi.org/10.3390/v14092025. Accessed 20 Mar 2025
Emert-Sedlak LA et al (2024) PROTAC-mediated degradation of HIV-1 Nef efficiently restores cell-surface CD4 and MHC-I expression and blocks HIV-1 replication. Cell Chem Biol 31:658–668.e14. https://doi.org/10.1016/j.chembiol.2024.02.004. Accessed 20 Mar 2025

EpiBiologics. https://www.epibiologics.com/. Accessed 20 Mar 2025

Evans LE et al (2019) Exploitation of antibiotic resistance as a novel drug target: development of a β-lactamase-activated antibacterial prodrug. J Med Chem 262:4411–4425. https://doi.org/10.1021/acs.jmedchem.8b01923. Accessed 20 Mar 2025

Expansion Therapeutics. https://www.expansionrx.com/. Accessed 20 Mar 2025

Fate Therapeutics. https://www.fatetherapeutics.com/. Accessed 20 Mar 2025

FDA Cell and Gene Therapy Approvals (2024). https://www.fda.gov/vaccines-blood-biologics/cellular-gene-therapy-products/approved-cellular-and-gene-therapy-products. Accessed 20 Mar 2025

Fralish Z et al (2024) The landscape of small-molecule prodrugs. Nat Rev Drug Discov 23:365–380. https://doi.org/10.1038/s41573-024-00914-7. Accessed 20 Mar 2025

Frangoul H et al (2024) Exagamglogene Autotemcel for severe sickle cell disease. N Engl J Med 390:1649–1662. https://doi.org/10.1056/nejmoa2309676. Accessed 20 Mar 2025

Garber K (2023) Drugging RNA. Nat Biotechnol 41:745–749. https://doi.org/10.1038/s41587-023-01790-z. Accessed 20 Mar 2025

Generation Bio. www.generationbio.com. Accessed 20 Mar 2025

Gilberg E, Bajorath J (2019) Recent progress in structure-based evaluation of compound promiscuity. ACS Omega 4:2758–2765. https://doi.org/10.1021/acsomega.8b03639. Accessed 20 Mar 2025

Gircha AI et al (2023) Hybrid quantum-classical machine learning for generative chemistry and drug design. Sci Rep 13:8250. https://doi.org/10.1038/s41598-023-32703-4. Accessed 20 Mar 2025

Glögl M et al (2024) Target-conditioned diffusion generates potent TNFR superfamily antagonists and agonists. Science 386:1154–1161. https://doi.org/10.1126/science.adp1779. Accessed 20 Mar 2025

Grigalunas M et al (2022) Chemical evolution of natural product structure. J Am Chem Soc 144:3314–3329. https://doi.org/10.1021/jacs.1c11270. Accessed 20 Mar 2025

Harlow EM, Adair JE (2024) Make gene therapies more available by manufacturing them in lower-income nations. Nature 631:502–504. https://doi.org/10.1038/d41586-024-02310-y. Accessed 20 Mar 2025

Hayday A et al (2024) Cancer immunotherapy by γδ T cells. Science 386:eabq7248. https://doi.org/10.1126/science.abq7248. Accessed 20 Mar 2025

He H et al (2024a) De novo generation of SARS-CoV-2 antibody CDRH3 with a pre-trained generative large language model. Nat Commun 15:6867. https://doi.org/10.1038/s41467-024-50903-y. Accessed 20 Mar 2025

He J et al (2024b) Biocatalytic C–H oxidation meets radical cross-coupling: simplifying complex piperidine synthesis. Science 386:1421–1427. https://doi.org/10.1126/science.adr9368. Accessed 20 Mar 2025

Hemgenix EMA data. https://www.ema.europa.eu/en/documents/product-information/hemgenix-epar-product-information_en.pdf. Accessed 20 Mar 2025

Herrera M et al (2024) Bispecific antibodies: advancing precision oncology. Trends Cancer 10:893–919. https://doi.org/10.1016/j.trecan.2024.07.002. Accessed 20 Mar 2025

Herzberg BO, Manji GA (2023) KRAS: druggable at last. Oncologist 28:283–286. https://doi.org/10.1093/oncolo/oyad014. Accessed 20 Mar 2025

Hou X et al (2021) Lipid nanoparticles for mRNA delivery. Nat Rev Mater 6:1078–1094. https://doi.org/10.1038/s41578-021-00358-0. Accessed 20 Mar 2025

Hughes SJ et al (2021) The rise and rise of protein degradation: opportunities and challenges ahead. Drug Discov Today 26:2889–2897. https://doi.org/10.1016/j.drudis.2021.08.006. Accessed 20 Mar 2025

Hussack et al (2011) Engineered single-domain antibodies with high protease resistance and thermal stability. PLoS One 6: e28218 https://doi.org/10.1371/journal.pone.0028218. Accessed 20 Mar 2025

iECURE. https://iecure.com/. Accessed 20 Mar 2025

Igawa T et al (2013) Engineered monoclonal antibody with novel antigen-sweeping activity *in vivo*. PLoS One 8:e63236. https://doi.org/10.1371/journal.pone.0063236. Accessed 20 Mar 2025

In8Bio. https://in8bio.com/. Accessed 20 Mar 2025

Interius BioTherapeutics. https://interiusbio.com/. Accessed 20 Mar 2025

Ishida T, Ciulli C (2021) E3 ligase ligands for PROTACs: how they were found and how to discover new ones. Perspectives 26:484–502. https://doi.org/10.1177/2472555220965528. Accessed 20 Mar 2025

Ivanenkov YA et al (2023) Chemistry42: an AI-driven platform for molecular design and optimization. J Chem Inf Model 63:695–701. https://doi.org/10.1021/acs.jcim.2c01191. Accessed 20 Mar 2025

Jadhav V et al (2024) RNA interference in the era of nucleic acid therapeutics. Nat Biotechnol 42: 394–405. https://doi.org/10.1038/s41587-023-02105-y. Accessed 20 Mar 2025

Jin B-k, Odongo S, Radwanska M, Magez S (2023) NANOBODIES®: a review of diagnostic and therapeutic applications. Int J Mol Sci 24:5994. https://doi.org/10.3390/ijms24065994. Accessed 20 Mar 2025

Johns DG et al (2023) Orally bioavailable macrocyclic peptide that inhibits binding of PCSK9 to the low density lipoprotein receptor. Circulation 148:144–158. https://doi.org/10.1161/CIRCULATIONAHA.122.063372. Accessed 20 Mar 2025

Kaiser J (2025) Gene editor may have cured infant of a deadly metabolic disorder. Science 387: 234–235. https://doi.org/10.1126/science.adv9833. Accessed 20 Mar 2025

Kelonia Therapeutics. https://keloniatx.com/. Accessed 20 Mar 2025

Kenakin T (2024) Know your molecule: pharmacological characterization of drug candidates to enhance efficacy and reduce late-stage attrition. Nat Rev Drug Discov 23:626–644. https://doi.org/10.1038/s41573-024-00958-9. Accessed 20 Mar 2025

Kingwell K (2023) Macrocycle drugs serve up new opportunities. Nat Rev Drug Discov 22:771–773. https://doi.org/10.1038/d41573-023-00152-3. Accessed 20 Mar 2025

Kirkeby A et al (2025) Pluripotent stem-cell-derived therapies in clinical trial: a 2025 update. Cell Stem Cell 32:329–331. https://doi.org/10.1016/j.stem.2025.01.003. Accessed 20 Mar 2025

Kissman EN et al (2024) Expanding chemistry through in vitro and in vivo biocatalysis. Nature 631: 37–48. https://doi.org/10.1038/s41586-024-07506-w. Accessed 20 Mar 2025

Klein J et al (2014) Yeast synthetic biology platform generates novel chemical structures as scaffolds for drug discovery. ACS Synth Biol 3:314–323. https://doi.org/10.1021/sb400177x. Accessed 20 Mar 2025

Klein C et al (2017) Cergutuzumab amunaleukin (CEA-IL2v), a CEA-targeted IL-2 variant-based immunocytokine for combination cancer immunotherapy: overcoming limitations of aldesleukin and conventional IL-2-based immunocytokines. Onco Targets Ther 6. https://doi.org/10.1080/2162402X.2016.1277306. Accessed 20 Mar 2025

Klein D et al (2021) Centyrin ligands for extrahepatic delivery of siRNA. Mol Ther 29:2053–2066. https://doi.org/10.1016/j.ymthe.2021.02.015. Accessed 20 Mar 2025

Kliegman M et al (2024) A roadmap for affordable genetic medicines. Nature 634:307–314. https://doi.org/10.1038/s41586-024-07800-7. Accessed 20 Mar 2025

Koeberl D et al (2024) Interim analyses of a first-in-human phase 1/2 mRNA trial for propionic acidaemia. Nature 628:872–877. https://doi.org/10.1038/s41586-024-07266-7. Accessed 20 Mar 2025

Komor A et al (2016) Programmable editing of a target base in genomic DNA without double-stranded DNA cleavage. Nature 533:420–424. https://doi.org/10.1038/nature17946. Accessed 20 Mar 2025

Korro Bio. https://www.korrobio.com/. Accessed 20 Mar 2025

Lauffer MC et al (2024) Possibilities and limitations of antisense oligonucleotide therapies for the treatment of monogenic disorders. Commun Med 4:6. https://doi.org/10.1038/s43856-023-00419-1. Accessed 20 Mar 2025

Leonard WJ, Lin JX (2023) Strategies to therapeutically modulate cytokine action. Nat Rev Drug Discov 22:827–854. https://doi.org/10.1038/s41573-023-00746-x. Accessed 20 Mar 2025

Lindley SR et al (2024) Ribozyme-activated mRNA trans-ligation enables large gene delivery to treat muscular dystrophies. Science 386:762–767. https://doi.org/10.1126/science.adp8179. Accessed 20 Mar 2025

Liu E et al (2020) Use of CAR-transduced natural killer cells in CD19-positive lymphoid tumors. N Engl J Med 382:545–553. https://doi.org/10.1056/nejmoa1910607. Accessed 20 Mar 2025

Lovering F et al (2009) Escape from flatland: increasing saturation as an approach to improving clinical success. J Med Chem 52:6752–6756. https://doi.org/10.1021/jm901241e. Accessed 20 Mar 2025

Lowe KL et al (2019) Novel TCR-based biologics: mobilising T cells to warm 'cold' tumours. Cancer Treat Rev 77:35–43. https://doi.org/10.1016/j.ctrv.2019.06.001. Accessed 20 Mar 2025

Lycia Therapeutics. https://lyciatx.com/. Accessed 20 Mar 2025

Mallapaty S (2024a) Stem cells reverse woman's diabetes – a world first. Nature 634:271–272. https://doi.org/10.1038/d41586-024-03129-3. Accessed 20 Mar 2025

Mallapaty S (2024b) Cutting-edge CAR-T cancer therapy is now made in India-at one-tenth the cost. Nature 627:709–710. https://doi.org/10.1038/d41586-024-00809-y. Accessed 20 Mar 2025

Maxion Therapeutics. https://www.maxiontherapeutics.com/. Accessed 20 Mar 2025

Mehr SHM et al (2020) A universal system for digitization and automatic execution of the chemical synthesis literature. Science 370:101–108. https://doi.org/10.1126/science.abc2986. Accessed 20 Mar 2025

Metkar M et al (2024) Tailor made: the art of therapeutic mRNA design. Nat Rev Drug Discov 23: 67–83. https://doi.org/10.1038/s41573-023-00827-x. Accessed 20 Mar 2025

Mitchell DC et al (2023) A proteome-wide atlas of drug mechanism of action. Nat Biotechnol 41: 845–857. https://doi.org/10.1038/s41587-022-01539-0. Accessed 20 Mar 2025

Moradi M et al (2023) The paths and challenges of "off-the-shelf" CAR-T cell therapy: an overview of clinical trials. Biomed Pharmacother 169:115888. https://doi.org/10.1016/j.biopha.2023.115888. Accessed 20 Mar 2025

Muhuri M et al (2022) Durability of transgene expression after rAAV gene therapy. Mol Ther 30: 1364–1380. https://doi.org/10.1016/j.ymthe.2022.03.004. Accessed 20 Mar 2025

Mullard A (2023) Setting GPCRs free. Nat Rev Drug Discov 22:347–348. https://doi.org/10.1038/d41573-023-00064-2. Accessed 20 Mar 2025

Mullard A (2024a) When can AI deliver the drug discovery hits? Nat Rev Drug Discov 23:159–161. https://doi.org/10.1038/d41573-024-00036-0. Accessed 20 Mar 2025

Mullard A (2024b) Extracellular targeted protein degrader removes antibodies in first test in humans. Nat Rev Drug Disc 23:483–486. https://doi.org/10.1038/d41573-024-00103-6. Accessed 20 Mar 2025

Mullard A (2024c) FDA approves gene therapy for metachromatic leukodystrophy, the tenth for a genetic disease and the priciest yet. Nat Rev Drug Discov 23:328. https://doi.org/10.1038/d41573-024-00056-w. Accessed 20 Mar 2025

Mullard A (2024d) RNA-editing drugs advance into clinical trials. Nat Rev Drug Discov 23:323–326. https://doi.org/10.1038/d41573-024-00070-y. Accessed 20 Mar 2025

Mullard A (2024e) RNA-rewriting candidate moves into the clinic. Nat Rev Drug Discov 23:407–409. https://doi.org/10.1038/d41573-024-00086-4. Accessed 20 Mar 2025

Mullard A (2024f) In vivo CAR T cells move into clinical trials. Nat Rev Drug Discov 23:727–730. https://doi.org/10.1038/d41573-024-00150-z. Accessed 20 Mar 2025

Müller FD et al (2024) CD19 CAR T-cell therapy in autoimmune disease – a case series with follow-up. N Engl J Med 390:687–700. https://doi.org/10.1056/NEJMoa2308917. Accessed 20 Mar 2025

Mullowney MW et al (2023) Artificial intelligence for natural product drug discovery. Nat Rev Drug Discov 22:895–916. https://doi.org/10.1038/s41573-023-00774-7. Accessed 20 Mar 2025

Myeloid Therapeutics. https://myeloidtx.com/. Accessed 20 Mar 2025

Nagra NS et al (2023) The company landscape for artificial intelligence in large-molecule drug discovery. Nat Rev Drug Disc 22:949–950. https://doi.org/10.1038/d41573-023-00139-0. Accessed 20 Mar 2025

Nanite. https://nanitebio.com/. Accessed 20 Mar 2025
Nathan P et al (2021) Overall survival benefit with tebentafusp in metastatic uveal melanoma. N Engl J Med 385:1196–1206. https://doi.org/10.1056/nejmoa2103485. Accessed 20 Mar 2025
News feature (2023) New guise ADCs couple antibodies to degraders. Nature Biotech 41:1359. https://doi.org/10.1038/s41587-023-02001-5. Accessed 20 Mar 2025
Nimjee SM et al (2017) Aptamers as therapeutics. Annu Rev Pharmacol Toxicol 57:61–79. https://doi.org/10.1146/annurev-pharmtox-010716-104558. Accessed 20 Mar 2025
Notin P et al (2024) Machine learning for functional protein design. Nat Biotechnol 42:216–228. https://doi.org/10.1038/s41587-024-02127-0. Accessed 20 Mar 2025
Nurix. https://www.nurixtx.com/. Accessed 20 Mar 2025
Obexer R et al (2024) Modern approaches to therapeutic oligonucleotide manufacturing. Science 384:6692. https://doi.org/10.1126/science.adl4015. Accessed 20 Mar 2025
Offensperger F et al (2024) Large-scale chemoproteomics expedites ligand discovery and predicts ligand behavior in cells. Science 384:eadk5864. https://doi.org/10.1126/science.adk5864. Accessed 20 Mar 2025
Orbital Therapeutics. https://www.orbitaltx.com/. Accessed 20 Mar 2025
Orna. https://www.ornatx.com/. Accessed 20 Mar 2025
Ostrem J et al (2013) K-Ras(G12C) inhibitors allosterically control GTP affinity and effector interactions. Nature 503:548–551. https://doi.org/10.1038/nature12796. Accessed 20 Mar 2025
Parakh S et al (2020) Radiolabeled antibodies for cancer imaging and therapy. Cancers 14:1454. https://doi.org/10.3390/cancers14061454. Accessed 20 Mar 2025
PASQAL. https://www.pasqal.com/. Accessed 20 Mar 2025
Patel S et al (2022) Results of a phase I trial of FT500, a first-in-class, off-the-shelf, iPSC–derived NK cell therapy combined with PD-1/PD-L1 checkpoint blockade therapy and IL-2 in patients with advanced solid tumors. J Immunother Cancer 10. https://doi.org/10.1136/jitc-2022-SITC2022.0755. Accessed 20 Mar 2025
Patkin JL et al (2021) An engineered IL-2 reprogrammed for anti-tumor therapy using a semi-synthetic organism. Nat Commun 12:4785. https://doi.org/10.1038/s41467-021-24987-9. Accessed 20 Mar 2025
Pedram K et al (2023) Design of a mucin-selective protease for targeted degradation of cancer-associated mucins. Nat Biotechnol 42:597–607. https://doi.org/10.1038/s41587-023-01840-6. Accessed 20 Mar 2025
Peterson AA, Liu DR (2023) Small-molecule discovery through DNA-encoded libraries. Nat Rev Drug Discov 22:699–722. https://doi.org/10.1038/s41573-023-00713-6. Accessed 20 Mar 2025
Pfeiffer LS, Stafforst T (2023) Precision RNA base editing with engineered and endogenous effectors. Nat Biotechnol 41:1526–1542. https://doi.org/10.1038/s41587-023-01927-0. Accessed 20 Mar 2025
Poletti V, Mavilio F (2021) Designing Lentiviral vectors for gene therapy of genetic diseases. Viruses 13:1526. https://doi.org/10.3390/v13081526. Accessed 20 Mar 2025
Popow J et al (2024) Targeting cancer with small-molecule pan-KRAS degraders. Science 385: 1338–1347. https://doi.org/10.1126/science.adm8684. Accessed 20 Mar 2025
Precision Biosciences. https://precisionbiosciences.com/. Accessed 20 Mar 2025
Prime Medicine. https://primemedicine.com/. Accessed 20 Mar 2025
Promsote W et al (2023) Trispecific antibody targeting HIV-1 and T cells activates and eliminates latently-infected cells in HIV/SHIV infections. Nat Commun 14:3719. https://doi.org/10.1038/s41467-023-39265-z. Accessed 20 Mar 2025
ProQR Therapeutics. https://www.proqr.com/. Accessed 20 Mar 2025
PTC Therapeutics. https://www.ptcbio.com/. Accessed 20 Mar 2025
Qubit Pharmaceuticals. https://www.qubit-pharmaceuticals.com/. Accessed 20 Mar 2025
Raouf YS (2024) Covalent inhibitors: to infinity and beyond. J Med Chem 67:10513–10516. https://doi.org/10.1021/acs.jmedchem.4c01308. Accessed 20 Mar 2025
RCSB PDB. https://www.rcsb.org/. Accessed 20 Mar 2025
Recode Therapeutics. https://recodetx.com/. Accessed 20 Mar 2025
Remix Therapeutics. https://www.remixtx.com/. Accessed 20 Mar 2025

Rentería-Gómez A, Gutierrez O (2024) Atom-swap chemistry speeds synthesis of compounds for drug discovery. Nature 631:30–31. https://doi.org/10.1038/d41586-024-02017-0. Accessed 20 Mar 2025

Rhodes CA, Pei D (2017) Bicyclic peptides as next-generation therapeutics. Chemistry 23:12690–12703. https://doi.org/10.1002/chem.201702117. Accessed 20 Mar 20252

Ribometrix. https://ribometrix.com/. Accessed 20 Mar 2025

Rivnay J et al (2024) Are implantable, living pharmacies within reach? Science 386:271–273. https://doi.org/10.1126/science.adq5182. Accessed 20 Mar 2025

Rollins ZA et al (2024) MolPROP: molecular property prediction with multimodal language and graph fusion. J Cheminform 16:56. https://doi.org/10.1186/s13321-024-00846-9. Accessed 20 Mar 2025

Rosenberg SA (2024) Lymphocytes as a living drug for cancer. Science 385:25–26. https://doi.org/10.1126/science.adp1130. Accessed 20 Mar 2025

Rouce RH, Porteus MH (2024) Cell and gene therapy accessibility, vol 385. Science, p 475. https://doi.org/10.1126/science.ads0252. Accessed 20 Mar 2025

Sadybekov AV, Katritch V (2023) Computational approaches streamlining drug discovery. Nature 616:673–685. https://doi.org/10.1038/s41586-023-05905-z. Accessed 20 Mar 2025

Saez-Ibañez AR et al (2024) The changing landscape of cancer cell therapies: clinical trials and real-world data. Nat Rev Drug Discov 23:736–737. https://doi.org/10.1038/d41573-024-00094-4. Accessed 20 Mar 2025

Sandeep SH et al (2024) Engineered polyspecific antibodies: a new frontier in the field of immunotherapeutics. Immunology 171:464–496. https://doi.org/10.1111/imm.13743. Accessed 20 Mar 2025

Santos PM, Butterfield LH (2018) Dendritic cell–based cancer vaccines. J Immunol 200:443–449. https://doi.org/10.4049/jimmunol.1701024. Accessed 20 Mar 2025

Schulze CJ et al (2023) Chemical remodeling of a cellular chaperone to target the active state of mutant KRAS. Science 381:794–799. https://doi.org/10.1126/science.adg9652. Accessed 20 Mar 2025

Service RF (2024) Compound interest. Science 384:6699–6953. https://doi.org/10.1126/science.adq6634. Accessed 20 Mar 2025

Shanker VR et al (2024) Unsupervised evolution of protein and antibody complexes with a structure-informed language model. Science 385:46–53. https://doi.org/10.1126/science.adk8946. Accessed 20 Mar 2025

Sheridan C (2023) Why gene therapies must go virus-free. NatBiotech 41:737–744. https://doi.org/10.1038/s41587-023-01824-6. Accessed 20 Mar 2025

Sheridan C (2024) Innovators want pills to treat sickle cell disease. Can they match gene therapy? Nat Biotechnol 42:347–350. https://doi.org/10.1038/s41587-024-02179-2. Accessed 20 Mar 2025

Šícho M et al (2023) DrugEx: deep learning models and tools for exploration of drug-like chemical space. J Chem Inf Model 63:3629–3636. https://doi.org/10.1021/acs.jcim.3c00434. Accessed 20 Mar 2025

Skyhawk Therapeutics. https://www.skyhawktx.com/. Accessed 20 Mar 2025

Sorbie JA, Diao T (2024) Reaction for coupling 3D molecular fragments expands the chemist's toolkit. Nature 634:551–553. https://doi.org/10.1038/d41586-024-03221-8. Accessed 20 Mar 2025

Sun Y et al (2024) In vivo editing of lung stem cells for durable gene correction in mice. Science 384:1196–1202. https://doi.org/10.1126/science.adk9428. Accessed 20 Mar 2025

Swingle KL et al (2025) Placenta-tropic VEGF mRNA lipid nanoparticles ameliorate murine pre-eclampsia. Nature 637:412–421. https://doi.org/10.1038/s41586-024-08291-2. Accessed 20 Mar 2025

Systimmune Inc. https://systimmune.com/. Accessed 20 Mar 2025

Tang Q, Khvorova A (2024) RNAi-based drug design: considerations and future directions. Nat Rev Drug Discov 23:341–364. https://doi.org/10.1038/s41573-024-00912-9. Accessed 20 Mar 2025

Tasfaout H et al (2024) Split intein-mediated protein trans-splicing to express large dystrophins. Nature 632:192–200. https://doi.org/10.1038/s41586-024-07710-8. Accessed 20 Mar 2025

Ting PY et al (2024) A molecular glue degrader of the WIZ transcription factor for fetal hemoglobin induction. Science 385:91–99. https://doi.org/10.1126/science.adk6129. Accessed 20 Mar 2025

Tong Y et al (2023) Programming inactive RNA-binding small molecules into bioactive degraders. Nature 618:169–179. https://doi.org/10.1038/s41586-023-06091-8. Accessed 20 Mar 2025

Tropsha A et al (2024) Integrating QSAR modelling and deep learning in drug discovery: the emergence of deep QSAR. Nat Rev Drug Discov 23:141–155. https://doi.org/10.1038/s41573-023-00832-0. Accessed 20 Mar 2025

Turzo SBA et al (2022) Applications of machine learning in computer-aided drug discovery. QRB Discov 3:1–16. https://doi.org/10.1017/qrd.2022.12. Accessed 20 Mar 2025

UCB. https://www.ucb.com/innovation/drug-discovery. Accessed 20 Mar 2025

Ueda A et al (2023) Toward the development of Epigenome editing-based therapeutics: potentials and challenges. Int J Mol Sci 24:4778. https://doi.org/10.3390/ijms24054778. Accessed 20 Mar 2025

Umoja Biopharma. https://www.umoja-biopharma.com/. Accessed 20 Mar 2025

Vakili MG et al (2025) Quantum-computing-enhanced algorithm unveils potential KRAS inhibitors. Nat Biotechnol. https://doi.org/10.1038/s41587-024-02526-3. Accessed 20 Mar 2025

Vertex Type I diabetes cell therapy. https://www.vrtx.com/our-science/pipeline/. Accessed 20 Mar 2025

Vico Therapeutics. https://vicotx.com/. Accessed 20 Mar 2025

Vogt M (2023) Exploring chemical space - generative models and their evaluation. Artif Intell Life Sci 3:100064. https://doi.org/10.1016/j.ailsci.2023.100064. Accessed 20 Mar 2025

Wang JH et al (2024) Adeno-associated virus as a delivery vector for gene therapy of human diseases. Sig Transduct Target Ther 9:78. https://doi.org/10.1038/s41392-024-01780-w. Accessed 20 Mar 2025

Wave Life Sciences. https://wavelifesciences.com/. Accessed 20 Mar 2025

Wearing ER et al (2024) Visible light–mediated aza Paternò–Büchi reaction of acyclic oximes and alkenes to azetidines. Science 384:1468–1476. https://doi.org/10.1126/science.adj6771. Accessed 20 Mar 2025

Whittaker MN, Musunuru K (2024) An epigenetic editor to silence genes, vol 384. Science, pp 1407–1408. https://doi.org/10.1126/science.adq3334. Accessed 20 Mar 2025

Willyard C (2024) Do cutting-edge CAR-T-cell therapies cause cancer? What the data say. Nature 629:22–24. https://doi.org/10.1038/d41586-024-01215-0. Accessed 20 Mar 2025

Wong F et al (2024) Discovery of a structural class of antibiotics with explainable deep learning. Nature 626:177–185. https://doi.org/10.1038/s41586-023-06887-8. Accessed 20 Mar 2025

Xie W et al (2023) Genomic medicines: the coming waves? Nat Rev Drug Discov 22:691–692. https://doi.org/10.1038/d41573-023-00098-6. Accessed 20 Mar 2025

Yin M-Y et al (2024) Self-regulating CAR-T cells modulate cytokine release syndrome in adoptive T-cell therapy. J Exp Med 221:e20221988. https://doi.org/10.1084/jem.20221988. Accessed 20 Mar 2025

Zhang M et al (2024) Catalytic asymmetric synthesis of meta benzene isosteres. Nature 633:90–95. https://doi.org/10.1038/s41586-024-07865-4. Accessed 20 Mar 2025

Zhong Y, Liu J (2024) Emerging roles of CAR-NK cell therapies in tumor immunotherapy: current status and future directions. Cell Death Discov 10:318. https://doi.org/10.1038/s41420-024-02077-1. Accessed 20 Mar 2025

Development and Implementation

7

7.1 Introduction

The main focus of this book is on the evolving landscape of drug target discovery and the interventions acting on those targets. The landscape is characterised by technical advances, both laboratory generated and computational, that can also be applied to the later preclinical and clinical stages of drug discovery. These stages are possibly the riskiest parts of the biopharmaceutical enterprise with high failure rates both before and after administration to human subjects. This chapter highlights some of the approaches being used to mitigate this high failure rate and concludes with a short coverage of drug delivery and diagnostics, key aspects of the implementation of approved medicines in clinical practice.

7.2 Preclinical Development

The components of the preclinical drug development phase from candidate selection to first time in humans has been outlined in Chap 2. This phase is a major hurdle for drug development as it has a failure rate as high as 90% or else provides candidates (particularly in oncology) with a high failure rate in later clinical trials. The problem is one of optimising the translational research required to bridge the gap between basic science and clinical development. A detailed analysis is provided by Seyhan (2019) who discusses the technical and cultural factors behind the slow progress in preclinical development and some of the ways in which this might be remedied (also discussed below).

7.2.1 Replacing Animals

The use of animals in biopharma R&D has always provoked controversy; the debate is polarised between those who see no need for animals given the state of current

E. D. Zanders, *Twenty-first Century Drug Discovery*,
https://doi.org/10.1007/978-3-031-93949-5_7

technology and those who see no alternative to at least some testing in species other than humans. Both sides are probably agreed on the fact that in an ideal world there would be no need use animals for *in vivo* evaluation of drug candidates, pharmacokinetics, safety pharmacology and toxicology.

The main issue in preclinical development is how well the data from animal models can be translated into humans and vice versa. It is well known that this translation depends upon the system of interest and that while some are predictive, others are not. This also occurs in reverse, with potential toxicities revealed in animals but absent in humans. The complexities occurring when comparing multiple animal species with human toxicology are illustrated by Pognan et al. (2023). The S1P1/5 receptor modulator Siponimod was shown to generate a high incidence of haemangiosarcoma in mice but not in rats; gene expression analysis in both drug-treated species showed equal activation of vascular endothelial cells (VECs) by gene expression profiling with only transient expression of mitotic genes and the cytokine PIGF2 in rats compared with a constant activation in mice. Human primary VECs behaved similarly to those of rats and not mice so Sipomimod was deemed safe enough to satisfy the regulators.

New Approach Methodology (NAM) is being developed to reduce or eliminate the need for animals in drug development and involves among other things, the creation of human tissues, *ex vivo* testing and computer modelling based on large quantities of real-world data. Whatever the scientific merits of any new approach to preclinical evaluation, there has to be formal acceptance by the regulatory authorities. Nelson et al. (2024) have covered this briefly from the point of view of the FDA.

Figure 7.1 gives an overview of the existing and evolving technologies being applied to preclinical development based in part on the summary of pharmaceutical industry practice in the area surveyed by Pognan et al. (2023).

7.2.1.1 Stem Cells and Organoids

The manipulation of human stem cells *in vitro* has provided differentiated cell lines that can be used for drug testing in a number of ways. For example, cells can be generated from individuals with and without a disease of interest providing a useful controlled screening model for testing interventions. For preclinical development, liver cells are central to drug metabolism and toxicology studies, so creating viable high throughput human liver cell cultures is important. Unfortunately human hepatocytes dedifferentiate in standard 2-D cultures which compromises long term testing. Induced pluripotent stem cells (iPS) theoretically provide an unlimited supply of human hepatocytes, but this may be time consuming and expensive. It is however likely to be a better option than relying solely on primary human cells or transformed HepG2 cells. Liver cell cultures are being formatted in different ways to maintain the differentiated state, retain CYP450 function, be cost effective and useable at high throughput. These formats include 3-D cultures, co-cultures with more than one cell type, and perfusion-based systems (Cox et al. 2020; Yang et al. 2023).

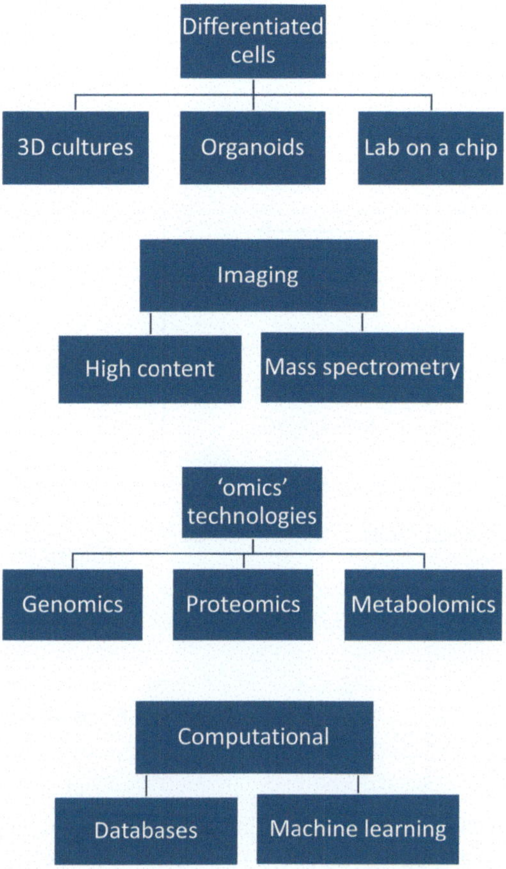

Fig. 7.1 Technology areas for preclinical development

Organoids derived from stem cells are the ultimate in modelling 3D organisation as they can reproduce at least some of the complexity of whole organs. Figure 7.2 shows a summary of organoid types generated from human ES cells or iPCs.

Organoids representing various tissues are now used routinely in biopharma R&D (Tang et al. 2022), but there will always be room for further development to align the models as closely as possible to the natural organ *in vivo*. One way in which this can be explored is to introduce more than one tissue type within the organoid. For example, Recaldin et al. (2024) have created human intestinal organoids populated with autologous tissue resident memory T lymphocytes. These intestinal immune-organoids reflect the natural situation in that the T-cell compartment surveys the intestinal epithelium and can undergo an inflammatory response. Many therapeutic drugs cause adverse events in patients with varying degrees of severity. This is illustrated by solitomab, a bispecific antibody targeting activated T-cells and the tumour antigen EpCAM whose development was terminated due to unacceptable intestinal inflammation. This problem could have been anticipated, at

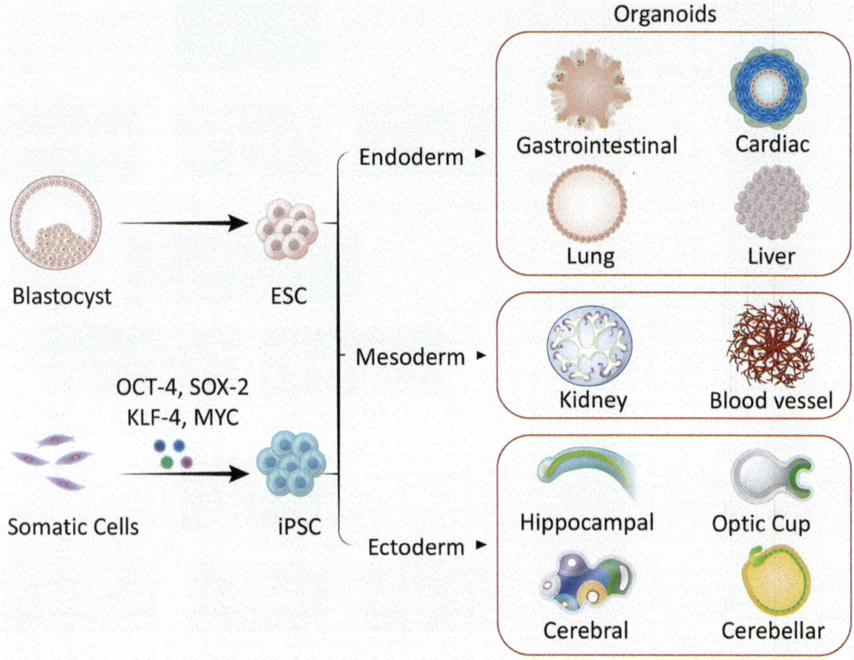

Fig. 7.2 Organoid types and origins from ES and iPS cells. (Figure taken from Tang et al. (2022) under CC-BY-4.0)

least in principle, by using the intestinal immune-organoid system to recreate the inflammatory response *ex vivo* prior to clinical development (Recaldin et al. 2024).

7.2.1.2 Microphysiology and Lab-on-a-Chip

In a sense, the previous sections on human cell and organoid models for drug development describe the components of the ultimate animal-free tests, namely living physiological systems accurately recreated in a controlled laboratory environment. The required microphysiology is realised through microfluidic devices, otherwise known as *lab-on-a-chip* or *organs-on-chips*. The various formats and applications of these devices to drug discovery and development have been reviewed by Ingber (2022). Figure 7.3 shows an artists' impression of the Microbrain Bioreactor Chip created as a concept by Vanderbilt University and collaborators (see VIIBRE 2025). The device is designed to reproduce the brain's microenvironment with neurons and blood brain barrier cells fed with a blood-carrying capillary. The lower layer contains an artificial choroid plexus (red) producing CSF and a venule (blue) to remove blood from the brain, along with cells forming the blood-CSF and CSF-brain barriers.

Current multi-organ chips are highly sophisticated; they have real potential for replacing *in vivo* tests with animals in the future but depend upon the quality of the individual cells and organoids used in the devices. There is also the important issue

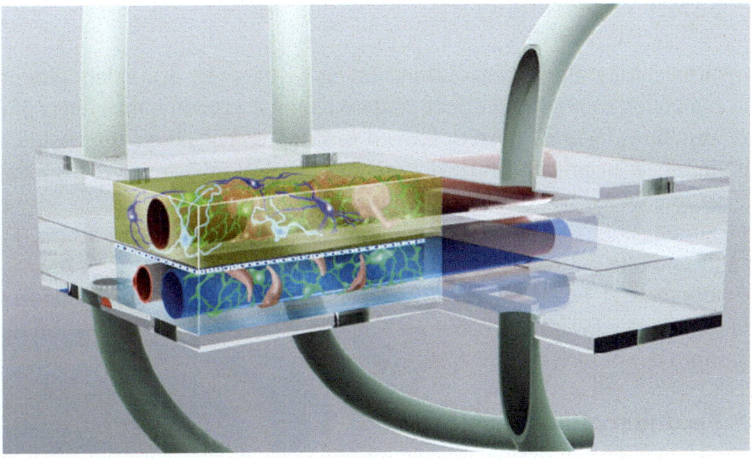

Fig. 7.3 Artist's impression of Microbrain Bioreactor Chip from VIIBRE created by Doyle D and Block F, Vanderbilt University. Details are in the main text

of implementation by drug developers and approval by the regulators, a far from trivial issue. Baran et al. (2022) have clearly laid this out in the form of a workshop report involving the FDA to discuss the use of microfluidic systems as drug development tools (DDTs). As defined by the FDA:

Context of use is the manner and purpose of use for a DDT; when FDA qualifies a biomarker, it is qualified for a specific context of use. The context of use statement should describe all of the elements characterizing this purpose and manner of use. The qualified context of use defines the boundaries with-in which the available data adequately justify use of the DDT.

Qualifying a liver-based microfluidic organ-on-a-chip model to predict risk for human drug-induced liver injury (DILI) would likely require a robust data package with responses from large test and validation sets that represent the many different aetiologies associated with this toxicity. The qualification package for internal (e.g., Sponsor) decision-making may be different from what is required for regulatory acceptance. The factors outlined above make defining simple guidelines for qualification around COUs for MPS or other exploratory *in vitro* models challenging (taken from Baran et al. (2022). In other words, the higher the level of regulation the more difficult it will be to implement some of the technically demanding packages. There is also the problem of how new types of drug such as cells and gene therapies should be evaluated as, unlike with small molecules, biopharma industry experience with these agents is limited. One of the conclusions from the workshop was that microphysiology devices should be based on animal models as well as humans in order to assess concordance between different species.

7.2.2 Safety Assessments

Safety pharmacology: the effects of drugs on body functions, secondary pharmacology: the evaluation of off-target effects and toxicology: assessing the effects of drugs on body structures have all been described in Chap 2. The above discussion about replacing animals in preclinical development is highly relevant to these areas. As well as being covered by formal regulations and recommendations (e.g. the ICH), there are organisations such as the Predictive Safety Testing Consortium (PSTC) (2025) and The International Consortium for Innovation and Quality in Pharmaceutical Development (IQ Consortium 2025). This section covers recent thinking about secondary pharmacology and toxicology and how these core development functions may be improved in the future.

7.2.2.1 Secondary Pharmacology

Pharmacological activity of drug-like small molecules beyond their intended targets is a fact of life in drug discovery (Brennan et al. 2024). Many small-molecule drugs interact with multiple protein targets, one example being the covalent binding drug aspirin, with 23 putative binding sites (Dai et al. (2016). In this case, the polypharmacology of aspirin is an advantage given its multiple beneficial effects, but more often, off-target-related toxicity is a major cause of attrition in drug discovery with at least nine drugs across a wide range of classes being withdrawn from the market or faced labelling restrictions. This is often due to inhibition of the KV11.1 (hERG) potassium channel associated with QT prolongation and potentially fatal arrhythmias. This vulnerability has been recognised for some time and is now part of the ICH guidelines (Chap 2). Ion channels of this type represent a very small part of the human proteome that may create problems with off-target binding; the numbers involved are challenging for the conventional secondary pharmacology approach of screening drug candidates against a panel of protein classes. Novel drug types also create uncertainty an example being targeted protein degraders whose possible degradation of unintended targets would be missed using conventional assays.

The discovery and evaluation of protein drug targets have been covered extensively in Chap. 5; the question here is how to translate research findings into a formal preclinical development process. Brennan et al. (2024) present recommendations for increasing secondary pharmacology screening panels as well as reporting procedures based on surveys of DruSafe, the Preclinical Safety arm of the IQ Consortium comprised of biopharma industry experts. Some of the recommended secondary screening target classes are shown in Table 7.1. GPCRs make up a significant proportion of marketed drugs, so their large representation in the table is to be expected. Few kinases have been used in the past however, but they do provide significant off-target liabilities often due to conserved orthosteric sites between different enzymes.

Table 7.1 Secondary screening targets from Brennan et al. (2024) with permission

Target class	Number in survey
Kinases	453
GPCRs	141
Ion channels	58
Transporters	16
Nuclear hormone receptors	14

7.2.2.2 Toxicology

Regulatory toxicity assessment prior to *first-in-human* clinical trials, requires good laboratory practice (GLP) studies *in vivo* to identify areas of concern, e.g. mechanistic toxicity via the primary drug target. Traditional *in vivo* assessment is being supplemented with investigative toxicology, a broad approach involving a range of *in vitro* and *in vivo* investigations including those described earlier. Investigative toxicology is discussed in detail by Pognan et al. (2023) and includes feedback from biopharma companies involved in this aspect of drug development.

7.2.3 Computational Approaches

7.2.3.1 Physiologically Based Pharmacokinetic (PBPK) Modelling

Computer analysis of data acquired during preclinical development is not new: For example, physiologically based pharmacokinetic (PBPK) modelling has been developed to quantify pharmacokinetic parameters (ADME) in human and animal subjects to support among other things, dosing estimations for first in human trials and safety pharmacology (Peters 2021). PBPK can cover the field of clinical assessment where a given individual can be differentiated from a specific virtual population according to age, sex, and other specific physiological features, in a sense comparing real patients with *virtual twins*. Tylutki et al. (2018) have applied this to precision dosing and personalized medicine using different responses of patients to the antidepressant amitriptyline. Population-based data indicate a highly variable PK of amitriptyline (AT) and its metabolite nortriptyline (NT) and adverse electrophysiological events especially in the situation of drug poisoning. The Tylutki et al. (2018) model accurately simulated the following:

The PK variability of orally administered AT and NT
Compared predictions versus clinically observed concentrations in differently characterized populations
Simulated PK of overdosed real individuals
Estimated individual active cardiac concentrations of AT and NT, and their variability in the population
Simulated the electrophysiological effect of AT and NT observed in clinical populations and overdosed patients

Developments in machine learning, highlighted in the next section, are being applied to Pharmacokinetic Modelling as reviewed (Chou and Lin 2023).

7.2.3.2 AI and Machine Learning

Random Forest

Some background to machine learning and deep learning has been given In Chap 3. The work of Ietswaart et al. (2020) on associating adverse drug reactions (ADRs) with *in vitro* pharmacology is an example of machine learning using decision trees based on random forests. Input parameters consisted of *in vitro* pharmacological data from 1329 marketed and withdrawn drugs plus adverse event reports compiled by the FDA's Adverse Event Reporting System (FAERS 2025). Data were processed using a random forest model to predict drug-ADR and target-ADR associations as described by Ietswaart et al. (2020). For a given (drug) input of AC_{50} values and ADR, the random forest model output, termed ADR probability, can be interpreted as the probability that the ADR is associated with the drug. By evaluating Gini importance scores[1] of model features, 221 target-ADR associations were detected; these co-occur in PubMed abstracts to a greater extent than expected by chance and include the well-established association of *in vitro* hERG binding with cardiac arrhythmias. Other associations include PDE3 with 40 ADRs which was not known previously thus supporting the potential utility of this machine learning approach in safety assessment.

Knowledge Graphs

Reproductive toxicology is a vital part of drug development and Evangelista et al. (2023) have used knowledge graphs to visualise interactions between specific birth defects, genes and small molecule drugs. Semantic relationships between objects are represented as triples: subject-predicate-object, for example, chicken-lays-eggs. A network consisting of nodes and edges is generated from a combination of triples and visualised as a knowledge graph. Graphs generated from multiple parameters can be used to visualise unexpected associations between different features. The inputs for the Evangelista et al. (2023) study were genes based on their association with pathogenicity, likelihood of small molecules crossing the placental barrier, prediction of birth defects, knowledge about known targets for particular drugs and the effects of small molecules on gene expression. An example knowledge graph is shown in Fig. 7.4 taken from the Reprotox-KG interface produced by Evangelista et al. (2023). In this case, valproic acid is known to cause neural tube defects and is linked on a knowledge graph to spina bifida. Compounds, genes and clinical defects are displayed in a way where connections can be made to inform toxicology studies.

Deep Learning

Deep neural networks (DNNs) whose basic principles were covered in Chap 3. are recent developments in machine learning and largely contribute to the widespread interest in AI. DNNs are beginning to be applied to different stages of the drug

[1] Also called Mean Decrease Impurity (MDI) score, the higher the score the more important the features.

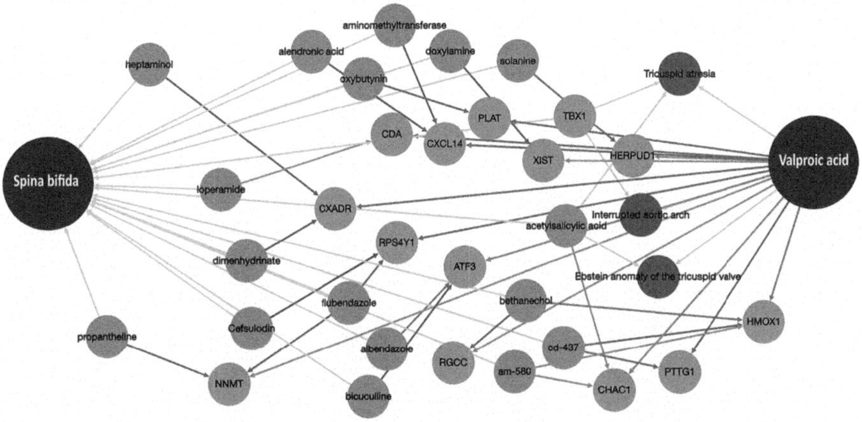

Fig. 7.4 Knowledge graph showing connections between valproic acid and spina bifida. (Taken from Reprotox-KG interface produced by Evangelista et al. (2023) under CC-BY-4.0)

discovery and development pipeline (Paul et al. (2021). For toxicology in particular, machine learning in the form of random forests and knowledge graphs have been highlighted above, but DNNs are also being developed for this purpose. Mayr et al. (2016) have created the DeepTox pipeline that outperformed other methods of computational toxicity predictions in the Tox21 Data Challenge. DeepTox computes a large number of chemical descriptors that are used as input to machine learning methods prior to model training and evaluation. The best candidates are combined to form ensembles. Lastly, the system predicts the toxicity of new compounds. Deep Learning learns abstract representations of the input data with higher levels of abstractions in higher layers. In image recognition for example, simple objects, such as edges and blobs, in lower layers are combined to abstract objects in higher layers. However in toxicology, it was not known how the data representations from Deep Learning could be interpreted. It could be shown that many hidden neurons represent previously known toxicophores i.e. those which have been handcrafted over decades by experts in the field. As well as these known toxicophores, it is reasonable to conclude that these include novel toxicophores that are latent in the data, this demonstrating the power of the deep learning approach (Mayr et al. (2016). Fig. 7.5 illustrates the emergence of known toxicophore features within in the first hidden layer of the DNN (rows 1 and 2, and a higher-level layer (row 3).

7.3 Clinical Development

A list of clinical trial phases was given in Chap 2. with some commentary about the phase 0 format for conducting pharmacokinetic and pharmacodynamic studies directly in human subjects using microdoses of experimental drugs. General comments about conducting clinical trials are given in ICH guideline E8

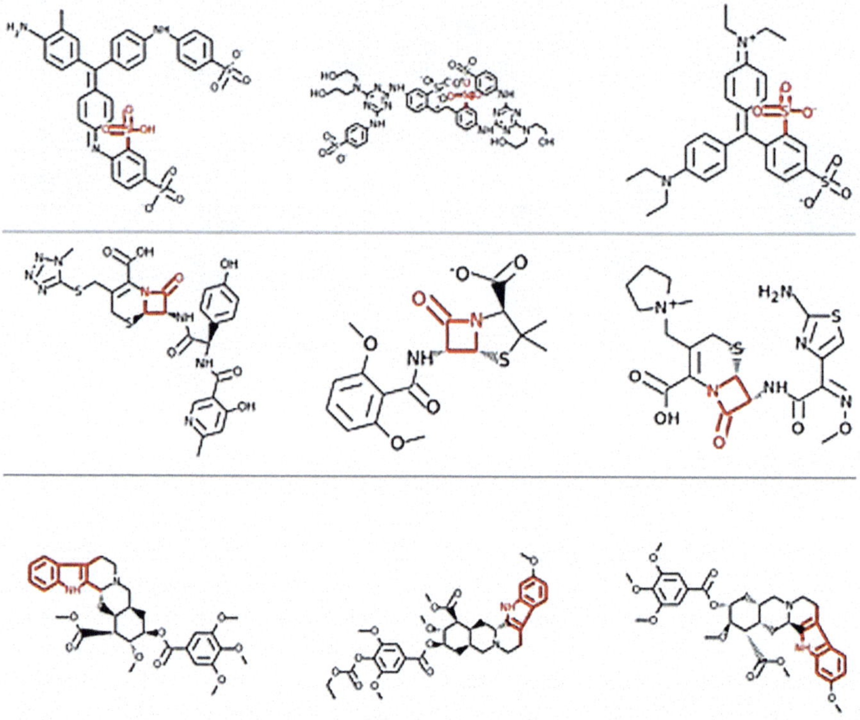

Fig. 7.5 Compounds with known toxicophores revealed in different layers of the DeepTox deep neural network. (Taken from Mayr et al. (2016) under CC-BY-4.0)

(R1 (2025) and more detailed information given on the ICH and other regulatory websites.

7.3.1 Clinical Trials

Clinical trials are an expensive and time-consuming part of the drug development process and can benefit from new technologies, such as machine learning and processes such as the acquisition of real world evidence.

7.3.1.1 Trial Design

Clinical trials can be observational or interventional, with the latter used to assess the effect of the medicine on patient outcomes, i.e. amelioration of disease and good tolerability. A breakdown of the trials listed in ClinicalTrials.gov (2025) as of 24th February 2025 is shown in Table 7.2. which conveys the amount of trial activity coordinated by the USA alone.

Trial designs are largely based on the randomised double blinded format that controls for bias in interpreting outcomes by patients and investigators. Trials of this

Table 7.2 Clinical Trials listed in ClinicalTrials.gov for 24th February 2025

Type	Number of trials
Interventional	403,521
Drug or biologic	201,536
Behavioural, other	153,171
Surgical procedure	42,192
Device	55,071
Observational	122,310
Expanded access	981

type are inflexible in that they must be run over a fixed time, perhaps lasting years, before being unblinded to possibly reveal no improvement in the patient group.

Adaptive trial designs are more flexible and can allow early termination of a particular intervention before moving on to alternatives. Adaptive trial designs are based on Bayesian statistics, i.e. not the problematic frequentist methods that rely on null hypothesis testing *p*-values and confidence intervals. The adaptive designs can provide gains in efficiency by potentially reducing the sample size (compared to a standard frequentist approach) and increasing allocations to more promising treatment arms, whilst maintaining high power.

One way of increasing the efficiency of drug development is to conduct rapid proof of concept trials in order to select the best treatment for further study. This approach was used during the COVID-19 pandemic to good effect when there was an urgent need to improve clinical outcomes which required the evaluation of unproven drug candidates. In the RECOVERY trial repurposed drugs such as dexamethasone and tocilizumab were shown to provide clinical benefit against the disease sufficiently rapidly to allow removal of patients from suboptimal treatment arms. Watson and Bandman (2023) have commented on how this principle can be applied to more clinical investigations to efficiently repurpose established treatments for new indications.

7.3.1.2 Increasing Efficiency Using Machine Learning

The complexity of clinical trials appears to be increasing over time (see Table 7.3) but often the tools used to implement and report them are inefficient and lacking in modern digital technology.

Hitesh et al. (2023) review the potential of machine learning for increasing clinical trial efficiency at the stages of patient recruitment, trial design and data processing.

Nearly a third of all phase 3 studies fail because of enrolment issues, and it is estimated that 86% of all trials do not reach recruitment schedules. Machine learning could help to select patients who are most likely to benefit from the proposed intervention. Liu et al. (2021) have devised the Trial Pathfinder system using oncology data from the database of lung cancer clinical trials (Flatiron Evidence Solutions 2025). Recruitment criteria such as thresholds for blood pressure and lymphocyte counts affect the rates of negative incidents among patients (hazard

Table 7.3 Increase in clinical trial complexity over time (from Hutson (2024) with permission

Category	2011–2015	2016–2020
Endpoints	13	22
Procedures	187	263
Eligibility criteria	30	30
Countries	9	15
Data points collected	929,203	3,453,133

Table 7.4 Selected AI/ML applications for clinical trials

Name	Applications	Reference
CliniDigest	Summary of clinical trial descriptions	White et al. (2023)
Panacea	All aspects foundation model	Lin et al. (2024)
Saama	Automation of protocol generation etc.	Saama (2025)
SEETrials	LLM to extract safety and efficacy information	IMO health (2025)
TrialGPT	Patient/trial selection	Jin et al. (2024)
Trial pathfinder	Patient/trial selection	Liu et al. (2021)

ratio). Trial Pathfinder could double the number of eligible patients without altering the hazard ratio thus opening trials up to more people with serious illnesses.

Large language models based on Chat GPT, such as TrialGPT, can be used to mine many thousands of clinical trial summaries and eliminate irrelevant trials. This is followed by criterion level annotation to accurately match patients to the appropriate trial and provide scores that outperform other methods (Jin et al. 2024).

Some examples of organisations developing clinical AI/ML are shown in Table 7.4.

While the intense activity in AI/ML is to be welcomed and will change the practice of clinical trials, there are caveats. Firstly, there is a need for robust data to input to the models. Much of the former may need to be converted from handwritten documents to machine readable forms. There may be bias in the selection of patients as the majority of clinical data (and genetic, see Chap 5.) are from people of European origin. Economic and other circumstances will also have an effect on the trial cohorts. There is also the much-noted fact that AI/ML algorithms are black boxes, where it is unclear how the results are generated. These problems will be solved over time, although the need for trained human observers will probably remain for the foreseeable future. If nothing else, this will be needed to avoid the legal complications that could arise if wrong clinical decisions are made as a result of total reliance on the algorithm's output.

7.3.1.3 Other Aspects of Clinical Trials

This section on clinical trials concludes with a brief discussion of how well their results are being disseminated and how they can be taken further in the form of postmarketing surveillance (PMS).

In the European Union, around 15% of clinical-trial results never get published or recorded. This can not only lead to research duplication but also create a false picture of a drug's potency and effectiveness, especially if negative results do not make it into the public sphere (Vespa 2024). Global reporting used to be consistently low, but new regulations like those from the FDA require timely submission of data and have improved the situation considerably. This is not the case for all countries, however. Part of the problem could be due to the time spent compiling reports, but this should become much less of an issue with the application of large language models etc.

The postmarketing surveillance (PMS) of a medicine occurs after its formal regulatory approval and will reflect the response of patients in the real world often having quite different characteristics to trial participants. In the USA, the FDA uses the Sentinel System to monitor curated clinical health care data from insurance claims plus some electronic health care records (EHRs). Its value has been enhanced by development of the Real World Evidence Data Enterprise that links EHR data to insurance claims for twenty-four million individuals across six participating data partners. Given the large amount of data that can be analysed, AI/ML models such as ChatGPT have potential for extracting useful drug safety information (Wang et al. 2023; Matheny et al. 2024) but with the usual caveats about accuracy and reliability. However, any improvements in PMS efficiency are to be welcomed as drug sponsors must conduct PMS studies for some years after approval; the Japanese MHLW for example, requires follow up of patients for 1–3 years after drug approval in order to monitor results in a large real-world population (Alexion 2024).

7.4 Drug Delivery

The biopharmaceutical industry has always favoured oral drug delivery for the obvious reasons of ease of manufacture, consistency of dosing and patient compliance. This favours small molecules at the expense of biotherapeutics and compounds with poor pharmacokinetics, such as rapid hepatic clearance. This picture is more complicated however, as discussed in an extensive review of drug delivery systems (DDS) by Baryakova et al. (2023). For example, poor adherence to medication is the most common reason for disparities observed between randomized clinical trials (RCTs) and real-world outcomes. Non-adherence is estimated to be around 50% for chronic illnesses and in the USA alone, is responsible for an estimated 125,000 deaths per annum, comparable with the number of deaths caused by colorectal cancer, breast cancer and prostate cancer combined. Poor adherence is also estimated to underlie $100–300 billion of avoidable health-care costs annually owing to wasted medicine, unnecessary diagnostic procedures and excessive health-care provider utilization.

From the perspective of pharmaceutical development, three quarters of phase II or phase III clinical trials were discontinued, in part, owing to efficacy and safety concerns due to poor pharmacokinetics and a narrow therapeutic window, respectively. Formulating a drug in a controlled-release system could overcome these problems and revive a failed development programme, thus validating the effort required to create a suitable DDS for each candidate.

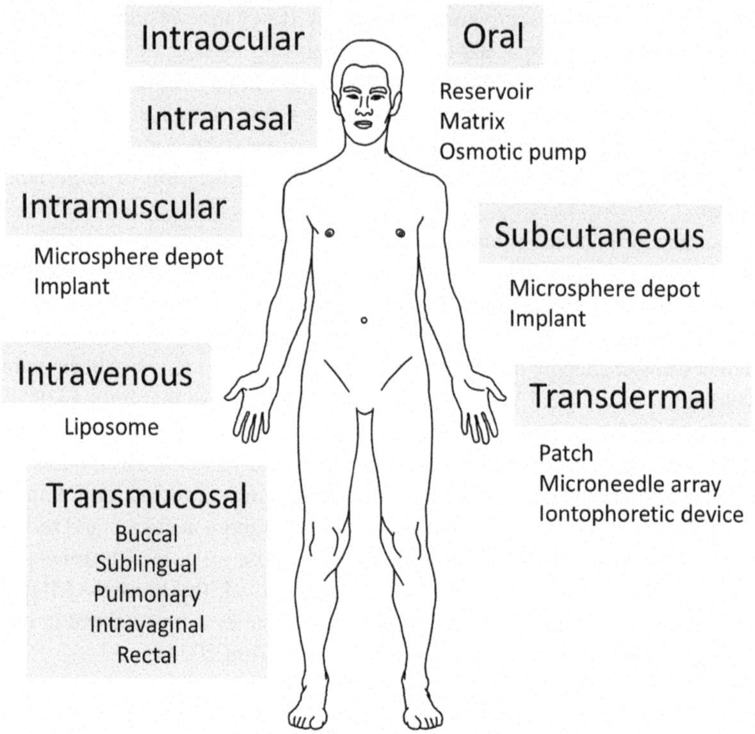

Fig. 7.6 Different routes of administration and related DDS adapted from Baryakova et al. (2023) with permission

The possible routes of drug administration to human subjects are illustrated in Fig. 7.6 along with some drug delivery systems (adapted from Baryakova et al. 2023).

Patient compliance with taking medicines and therefore gaining therapeutic benefit can be affected by factors such as:

The need for regular repeat dosing
Long time lag between dosing and benefit
Needle phobia

Current DDS systems include controlled release formulations used for different anatomical sites. Examples include intramuscular depots, delivery transdermally, or intranasal delivery for rapid onset of drug action in the CNS. Alternatives to hypodermic needles have been developed which is particularly important for vaccination campaigns. Microarray needle patches deliver drugs to the upper dermis avoiding pain receptors that reside further inside the skin (see Al-Japairai et al. 2020 for a review of the technology). Targeted protein and nucleic acid delivery via liposomes and viral vectors has been covered already in Chap 6. However, novel approaches are in early stages of development like the system for injecting proteins into target cells described by Kreitz et al. (2023). The contractile injection systems (CISs) used by bacteria for intercellular communication can be engineered to link a protein payload for injection into bacterial target cells as a proof of concept. Using AlfaFold to model protein domains it was possible to change the tropism of the CIS construct towards human cells thereby creating a potential therapeutic and delivery combination.

7.4.1 Oral Biotherapeutics Delivery

Apart from improving on existing DDS approaches, a major future objective will involve oral or nasal delivery of proteins and possibly nucleic acid drugs. The obvious barrier to oral delivery is the highly acidic stomach environment plus tight mucosal barriers that slow entry into the circulation. Oral and nasal insulin formulations have been introduced but were commercially unsuccessful because of efficacy and side effect issues (Baryakova et al. 2023). Small ingestible devices are being developed such as the *self-orientating millimetre-scale applicator* (SOMA) with a design based on a self-righting leopard tortoise that injects insulin through the stomach wall after swallowing. The drug payload is limited however, and other devices are being researched, such as the microjet delivery system inspired by the jetting devices used by Cephalopods for locomotion (Arrick et al. 2024). Jet injection devices can be tailored for different gastrointestinal segments including the stomach and small intestine and shown to deliver insulin, a GLP1 analogue and siRNA in large animal models and achieve exposure levels similar to those obtained with subcutaneous delivery.

Since monoclonal antibodies and engineered derivatives are likely to be in use for many years, there is an incentive to deliver them orally despite the technical difficulties involved. The small injection devices described above will no doubt contribute, but it is also possible to create formulations that deliver intact antibodies to the colon without degradation in the stomach and small intestine (Yadav et al. 2023).

Finally, engineered microorganisms have potential as drug delivery systems (reviewed by Charbonneau et al. 2020). Recently, *Lactobacillus plantarum* WCFS1 (Lp) has been used to transport drug payloads into the brains of mice *in vivo*. Lp is commensal bacterium with high affinity for the olfactory epithelium and can be engineered with specific payloads for release and migration to the brain (Shen et al. 2025). Proof of concept studies have shown successful delivery of appetite suppressing hormones to the mouse brain with resulting *in vivo* activity.

7.5 Diagnostics

This final chapter concludes with a brief coverage of diagnostics. It is well accepted that early diagnosis of many diseases can save patients from later morbidity and possible mortality, as well as savings in healthcare costs through avoidance of long term and expensive therapies.

The so-called five pillars of diagnostics as listed by the NHS in the United Kingdom are:

Imaging
Endoscopy
Pathology
Physiological Measurement
Genomics

A major diagnostic objective is to rapidly test and deliver results, notably with cardiovascular events and also in situations where newborns with a genetic condition may require rapid assessment and treatment planning. Advances are being made in all areas, particularly with the application of machine learning to process imaging and other data. The instrumentation for generating the data to be analysed is being continually improved, for example with ultra rapid DNA sequencing for diagnosis, something that has been achieved in just over 7 hours (Gorzynski et al. 2022).
For cancer diagnosis, tissue samples offer the most accurate and informative data; however, surrogate markers in the form of circulating tumour DNA (ctDNA) or other material from liquid biopsies have real potential for early diagnosis and treatment monitoring (Connal et al. 2023).

Diagnosis using plasma protein biomarkers is well established but not always reliable, something that has affected the Alzheimer's field for many years. The gold standard diagnostic tool for Alzheimer's is imaging using positron emission tomography (PET) scanning) to detect amyloid and tau proteins in the brain. Cheaper and easier alternatives are essential, particularly for managing clinical trials where it can be difficult to ensure that subjects actually have the disease and not another type of dementia. Sampling cerebrospinal fluid (CSF) is one option, although quite invasive; blood plasma is ideal if it is possible to analyse the very small quantity of brain proteins mixed with highly abundant serum proteins. Biomarkers such as phosphorylated tau (p-tau 217) and glial fibrillary acidic protein (GFAP) are proving to be useful predictors, along with new candidates as summarised by Abbott (2024).

Lastly, mention should be made of the potential for wearable devices to provide not only heart rate and other biometric data, but also body fluid analysis for diagnostics. This future area of real-world monitoring has the potential to greatly improve aspects of drug development as well as clinical practice and preventative medicine (Brasier et al. 2024).

References

Abbott A (2024) Blood tests could soon predict your risk of Alzheimer's. Nature 632:243–245. https://doi.org/10.1038/d41586-024-02535-x. Accessed 20 Mar 2025

Alexion (2024) What comes after clinical trials? https://www.nature.com/articles/d42473-023-00335-x. Accessed 20 Mar 2025

Al-Japairai KAS et al (2020) Current trends in polymer microneedle for transdermal drug delivery. Int J Pharm 587:119673. https://doi.org/10.1016/j.ijpharm.2020.119673. Accessed 20 Mar 2025

Arrick G et al (2024) Cephalopod-inspired jetting devices for gastrointestinal drug delivery. Nature 636:481–487. https://doi.org/10.1038/s41586-024-08202-5. Accessed 20 Mar 2025

Baran SW et al (2022) Perspectives on the evaluation and adoption of complex in vitro models in drug development: workshop with the FDA and the pharmaceutical industry (IQ MPS affiliate). ALTEX 39:297–314. https://doi.org/10.14573/altex.2112203. Accessed 20 Mar 2025

Baryakova TH et al (2023) Overcoming barriers to patient adherence: the case for developing innovative drug delivery systems. Nat Rev Drug Discov 22:387–409. https://doi.org/10.1038/s41573-023-00670-0. Accessed 20 Mar 2025

Brasier N et al (2024) Applied body-fluid analysis by wearable devices. Nature 636:57–68. https://doi.org/10.1038/s41586-024-08249-4. Accessed 20 Mar 2025

Brennan RJ et al (2024) The state of the art in secondary pharmacology and its impact on the safety of new medicines. Nat Rev Drug Discov 23:525–545. https://doi.org/10.1038/s41573-024-00942-3. Accessed 20 Mar 2025

Charbonneau MR et al (2020) Developing a new class of engineered live bacterial therapeutics to treat human diseases. Nat Commun 11:1738. https://doi.org/10.1038/s41467-020-15508-1. Accessed 20 Mar 2025

Chou W-C, Lin Z (2023) Machine learning and artificial intelligence in physiologically based pharmacokinetic modelling. Toxicol Sci 191:1–14. https://doi.org/10.1093/toxsci/kfac101. Accessed 20 Mar 2025

ClinicalTrials.gov. https://clinicaltrials.gov/. Accessed 20 Mar 2025

Connal S et al (2023) Liquid biopsies: the future of cancer early detection. J Transl Med 21:118. https://doi.org/10.1186/s12967-023-03960-8. Accessed 20 Mar 2025

Cox C et al (2020) Current perspective: 3D spheroid models utilizing human-based cells for investigating metabolism-dependent drug-induced liver injury. Front Med Technol 2. https://doi.org/10.3389/fmedt.2020.611913. Accessed 20 Mar 2025

Dai S et al (2016) Proteome-wide prediction of targets for aspirin: new insight into the molecular mechanism of aspirin. PeerJ 4:e1791. https://doi.org/10.7717/peerj.1791. Accessed 20 Mar 2025

Evangelista J et al (2023) Toxicology knowledge graph for structural birth defects. Commun Med 3: 98. https://doi.org/10.1038/s43856-023-00329-2. Accessed 20 Mar 2025

FAERS. https://www.fda.gov/drugs/surveillance/fdas-adverse-event-reporting-system-faers Accessed 20 Mar 2025

Flatiron Evidence Solutions. https://flatiron.com/real-world-evidence Accessed 20 Mar 2025

Gorzynski JE et al (2022) Ultrarapid Nanopore genome sequencing in a critical care setting. N Engl J Med 386:700–702. https://doi.org/10.1056/NEJMc2112090. Accessed 20 Mar 2025

Hitesh C et al (2023) Revolutionizing clinical trials: the role of AI in accelerating medical breakthroughs. Int J Surg 109:4211–4220. https://doi.org/10.1097/js9.0000000000000705. Accessed 20 Mar 2025

Hutson M (2024) How AI is being used to accelerate clinical trials. Nature 627(8003):S2–S5. https://doi.org/10.1038/d41586-024-00753-x. Accessed 20 Mar 2025

ICH guideline E8 (R1). https://www.ema.europa.eu/en/documents/regulatory-procedural-guideline/ich-guideline-e8-r1-general-considerations-clinical-studies_en.pdf. Accessed 20 Mar 2025

Ietswaart R et al (2020) Machine learning guided association of adverse drug reactions with in vitro target-based pharmacology. EBioMedicine 57:102837. https://doi.org/10.1016/j.ebiom.2020.102837. Accessed 20 Mar 2025

IMO Health. https://www.imohealth.com/. Accessed 20 Mar 2025

Ingber DE (2022) Human organs-on-chips for disease modelling, drug development and personalized medicine. Nat Rev Genet 23:467–491. https://doi.org/10.1038/s41576-022-00466-9. Accessed 20 Mar 2025

IQ Consortium. https://iqconsortium.org/. Accessed 20 Mar 2025

Jin Q et al (2024) Matching patients to clinical trials with large language models. Nat Commun 15:9074. https://doi.org/10.1038/s41467-024-53081-z. Accessed 20 Mar 2025

Kreitz J et al (2023) Programmable protein delivery with a bacterial contractile injection system. Nature 616:357–364. https://doi.org/10.1038/s41586-023-05870-7. Accessed 20 Mar 2025

Lin J et al (2024) Panacea: A foundation model for clinical trial search, summarization, design, and recruitment. medRxiv. https://doi.org/10.1101/2024.06.26.24309548. Accessed 20 Mar 2025

Liu R et al (2021) Evaluating eligibility criteria of oncology trials using real-world data and AI. Nature 592:629–633. https://doi.org/10.1038/s41586-021-03430-5. Accessed 20 Mar 2025

Matheny ME et al (2024) Enhancing postmarketing surveillance of medical products with large language models. JAMA Netw Open 7:e2428276. https://doi.org/10.1001/jamanetworkopen.2024.28276. Accessed 20 Mar 2025

Mayr A et al (2016) DeepTox: toxicity prediction using deep learning. Front Environ Sci 3:80. https://doi.org/10.3389/fenvs.2015.00080. Accessed 20 Mar 2025

Nelson CP et al (2024) Advancing alternative methods to reduce animal testing. Science 386:724–726. https://doi.org/10.1126/science.adg6228. Accessed 20 Mar 2025

Paul D et al (2021) Artificial intelligence in drug discovery and development. Drug Discov Today 26:80–93. https://doi.org/10.1016/j.drudis.2020.10.010. Accessed 20 Mar 2025

Peters SA (2021) Physiologically based pharmacokinetic (PBPK) modeling and simulations: principles, methods, and applications in the pharmaceutical industry. Wiley, New Jersey. https://doi.org/10.1002/9781119497813. Accessed 20 Mar 2025

Pognan F et al (2023) The evolving role of investigative toxicology in the pharmaceutical industry. Nat Rev Drug Discov 22:317–335. https://doi.org/10.1038/s41573-022-00633-x. Accessed 20 Mar 2025

Predictive Safety Testing Consortium (PSTC). https://c-path.org/program/predictive-safety-testing-consortium/. Accessed 20 Mar 2025

Recaldin T et al (2024) Human organoids with an autologous tissue-resident immune compartment. Nature 633:165–173. https://doi.org/10.1038/s41586-024-07791-5. Accessed 20 Mar 2025

Saama. https://www.saama.com/. Accessed 20 Mar 2025

Seyhan AA (2019) Lost in translation: the valley of death across preclinical and clinical divide-identification of problems and overcoming obstacles. Transl Med Commun 4(18). https://doi.org/10.1186/s41231-019-0050-7. Accessed 20 Mar 2025

Shen H et al (2025) Engineered commensals for targeted nose-to-brain drug delivery. Cell 188:1545–1562.e16. https://doi.org/10.1016/j.cell.2025.01.017. Accessed 20 Mar 2025

Tang XY et al (2022) Human organoids in basic research and clinical applications. Sig Transduct Target Ther 7:168. https://doi.org/10.1038/s41392-022-01024-9. Accessed 20 Mar 2025

Tylutki Z et al (2018) Physiologically based pharmacokinetic-quantitative systems toxicology and safety (PBPK-QSTS) modeling approach applied to predict the variability of amitriptyline pharmacokinetics and cardiac safety in populations and in individuals. J Pharmacokinet Pharmacodyn 45:663–677. https://doi.org/10.1007/s10928-018-9597-6. Accessed 20 Mar 2025

Vespa I (2024) Clinical-trial reporting is on the rise. Nature 634:S21–S23. https://doi.org/10.1038/d41586-024-03317-1. Accessed 20 Mar 2025

VIIBRE. https://www.vanderbilt.edu/viibre/. Accessed 20 Mar 2025

Wang H et al (2023) Future of ChatGPT in pharmacovigilance. Drug Saf 46:711–713. https://doi.org/10.1007/s40264-023-01315-2. Accessed 20 Mar 2025

Watson JMS, Bandman O (2023) The potential of innovative trial design for efficiently evaluating repurposed drugs. Nat Rev Drug Discov 22:681–682. https://doi.org/10.1038/d41573-023-00129-2. Accessed 20 Mar 2025

White R et al (2023) CliniDigest: a case study in large language model based large-scale summarization of clinical trial descriptions. In: Association for computing machinery ACM international conference proceeding series, pp 396–402. https://doi.org/10.1145/3582515.3609559. Accessed 20 Mar 2025

Yadav V et al (2023) Colon targeted delivery and stabilization of monoclonal antibodies for local treatment of inflammatory bowel diseases. Inflammatory Bowel Diseases 29:S3–S4. https://doi.org/10.1093/ibd/izac247.008. Accessed 20 Mar 2025

Yang S et al (2023) Liver three-dimensional cellular models for high-throughput chemical testing. Cell Rep Methods 3:100432. https://doi.org/10.1016/j.crmeth.2023.100432. Accessed 20 Mar 2025

Index

E. D. Zanders, *Twenty-first Century Drug Discovery*,
https://doi.org/10.1007/978-3-031-93949-5

The manufacturer's authorised representative in the EU is Springer
Nature Customer Service Centre GmbH, Europaplatz 3, 69115 Heidelberg,
Germany. If you have any concerns regarding our products, please
contact ProductSafety@springernature.com

Printed and bound by CPI Group (UK) Ltd, Croydon, CR0 4YY
05/10/2026
02228129-0003